Glencoe McGraw-Hill

Algebra 2

Study Guide
and Intervention
Workbook

Mc Graw Hill **Glencoe**

New York, New York   Columbus, Ohio   Chicago, Illinois   Woodland Hills, California

To the Student This *Study Guide and Intervention Workbook* gives you additional examples and problems for the concept exercises in each lesson. The exercises are designed to aid your study of mathematics by reinforcing important mathematical skills needed to succeed in the everyday world. The materials are organized by chapter and lesson, with two *Study Guide and Intervention* worksheets for every lesson in *Glencoe Algebra 2*.

Always keep your workbook handy. Along with your textbook, daily homework, and class notes, the completed *Study Guide and Intervention Workbook* can help you review for quizzes and tests.

To the Teacher These worksheets are the same as those found in the Chapter Resource Masters for *Glencoe Algebra 2*. The answers to these worksheets are available at the end of each Chapter Resource Masters booklet as well as in your Teacher Wraparound Edition interleaf pages.

The McGraw·Hill Companies

Copyright © by The McGraw-Hill Companies, Inc. All rights reserved.
Except as permitted under the United States Copyright Act, no part of this publication may be reproduced or distributed in any form or by any means, or stored in a database or retrieval system, without prior written permission of the publisher.

Send all inquiries to:
Glencoe/McGraw-Hill
8787 Orion Place
Columbus, OH 43240

ISBN: 978-0-07-879055-3
MHID: 0-07-879055-7   *Study Guide and Intervention Workbook, Algebra 2*

Printed in the United States of America

3 4 5 6 7 8 9 10 009 14 13 12 11 10 09 08

Contents

Lesson/Title	Page

Lesson/Title	Page

1-1 Study Guide and Intervention

Expressions and Formulas

Order of Operations

Order of Operations	1. Simplify the expressions inside grouping symbols. 2. Evaluate all powers. 3. Do all multiplications and divisions from left to right. 4. Do all additions and subtractions from left to right.

Example 1

Evaluate $[18 - (6 + 4)] \div 2$.

$$[18 - (6 + 4)] \div 2 = [18 - 10] \div 2$$
$$= 8 \div 2$$
$$= 4$$

Example 2

Evaluate $3x^2 + x(y - 5)$ if $x = 3$ and $y = 0.5$.

Replace each variable with the given value.
$$3x^2 + x(y - 5) = 3 \cdot (3)^2 + 3(0.5 - 5)$$
$$= 3 \cdot (9) + 3(-4.5)$$
$$= 27 - 13.5$$
$$= 13.5$$

Exercises

Find the value of each expression.

1. $14 + (6 \div 2)$

2. $11 - (3 + 2)^2$

3. $2 + (4 - 2)^3 - 6$

4. $9(3^2 + 6)$

5. $(5 + 2^3)^2 - 5^2$

6. $5^2 + \frac{1}{4} + 18 \div 2$

7. $\dfrac{16 + 2^3 \div 4}{1 - 2^2}$

8. $(7 - 3^2)^2 + 6^2$

9. $20 \div 2^2 + 6$

10. $12 + 6 \div 3 - 2(4)$

11. $14 \div (8 - 20 \div 2)$

12. $6(7) + 4 \div 4 - 5$

13. $8(4^2 \div 8 - 32)$

14. $\dfrac{6 + 4 \div 2}{4 \div 6 - 1}$

15. $\dfrac{6 + 9 \div 3 + 15}{8 - 2}$

Evaluate each expression if $a = 8.2$, $b = -3$, $c = 4$, and $d = -\dfrac{1}{2}$.

16. $\dfrac{ab}{d}$

17. $5(6c - 8b + 10d)$

18. $\dfrac{c^2 - 1}{b - d}$

19. $ac - bd$

20. $(b - c)^2 + 4a$

21. $\dfrac{a}{d} + 6b - 5c$

22. $3\left(\dfrac{c}{d}\right) - b$

23. $cd + \dfrac{b}{d}$

24. $d(a + c)$

25. $a + b \div c$

26. $b - c + 4 \div d$

27. $\dfrac{a}{b + c} - d$

1-1 Study Guide and Intervention (continued)

Expressions and Formulas

Formulas A **formula** is a mathematical sentence that uses variables to express the relationship between certain quantities. If you know the value of every variable except one in a formula, you can use substitution and the order of operations to find the value of the unknown variable.

Example To calculate the number of reams of paper needed to print n copies of a booklet that is p pages long, you can use the formula $r = \dfrac{np}{500}$, where r is the number of reams needed. How many reams of paper must you buy to print 172 copies of a 25-page booklet?

Substitute $n = 172$ and $p = 25$ into the formula $r = \dfrac{np}{500}$.

$$r = \frac{(172)(25)}{500}$$
$$= \frac{43{,}000}{500}$$
$$= 8.6$$

You cannot buy 8.6 reams of paper. You will need to buy 9 reams to print 172 copies.

Exercises

For Exercises 1–3, use the following information.

For a science experiment, Sarah counts the number of breaths needed for her to blow up a beach ball. She will then find the volume of the beach ball in cubic centimeters and divide by the number of breaths to find the average volume of air per breath.

1. Her beach ball has a radius of 9 inches. First she converts the radius to centimeters using the formula $C = 2.54I$, where C is a length in centimeters and I is the same length in inches. How many centimeters are there in 9 inches?

2. The volume of a sphere is given by the formula $V = \dfrac{4}{3}\pi r^3$, where V is the volume of the sphere and r is its radius. What is the volume of the beach ball in cubic centimeters? (Use 3.14 for π.)

3. Sarah takes 40 breaths to blow up the beach ball. What is the average volume of air per breath?

4. A person's basal metabolic rate (or BMR) is the number of calories needed to support his or her bodily functions for one day. The BMR of an 80-year-old man is given by the formula $BMR = 12w - (0.02)(6)12w$, where w is the man's weight in pounds. What is the BMR of an 80-year-old man who weighs 170 pounds?

1-2 Study Guide and Intervention

Properties of Real Numbers

Real Numbers All real numbers can be classified as either rational or irrational. The set of rational numbers includes several subsets: natural numbers, whole numbers, and integers.

R	real numbers	{all rationals and irrationals}
Q	rational numbers	{all numbers that can be represented in the form $\frac{m}{n}$, where m and n are integers and n is not equal to 0}
I	irrational numbers	{all nonterminating, nonrepeating decimals}
N	natural numbers	{1, 2, 3, 4, 5, 6, 7, 8, 9, ...}
W	whole numbers	{0, 1, 2, 3, 4, 5, 6, 7, 8, ...}
Z	integers	{..., −3, −2, −1, 0, 1, 2, 3, ...}

Example Name the sets of numbers to which each number belongs.

a. $-\frac{11}{3}$ rationals (Q), reals (R)

b. $\sqrt{25}$

$\sqrt{25} = 5$ naturals (N), wholes (W), integers (Z), rationals (Q), reals (R)

Exercises

Name the sets of numbers to which each number belongs.

1. $\frac{6}{7}$

2. $-\sqrt{81}$

3. 0

4. 192.0005

5. 73

6. $34\frac{1}{2}$

7. $\frac{\sqrt{36}}{9}$

8. 26.1

9. π

10. $\frac{15}{3}$

11. $-4.\overline{17}$

12. $\frac{\sqrt{25}}{5}$

13. −1

14. $\sqrt{42}$

15. −11.2

16. $-\frac{8}{13}$

17. $\frac{\sqrt{5}}{2}$

18. $33.\overline{3}$

19. 894,000

20. −0.02

1-2 Study Guide and Intervention (continued)

Properties of Real Numbers

Properties of Real Numbers

Real Number Properties		
For any real numbers *a*, *b*, and *c*		
Property	**Addition**	**Multiplication**
Commutative	$a + b = b + a$	$a \cdot b = b \cdot a$
Associative	$(a + b) + c = a + (b + c)$	$(a \cdot b) \cdot c = a \cdot (b \cdot c)$
Identity	$a + 0 = a = 0 + a$	$a \cdot 1 = a = 1 \cdot a$
Inverse	$a + (-a) = 0 = (-a) + a$	If *a* is not zero, then $a \cdot \frac{1}{a} = 1 = \frac{1}{a} \cdot a$.
Distributive	$a(b + c) = ab + ac$ and $(b + c)a = ba + ca$	

Example Simplify $9x + 3y + 12y - 0.9x$.

$$9x + 3y + 12y - 0.9x = 9x + (-0.9x) + 3y + 12y \qquad \text{Commutative Property (+)}$$
$$= (9 + (-0.9))x + (3 + 12)y \qquad \text{Distributive Property}$$
$$= 8.1x + 15y \qquad \text{Simplify.}$$

Exercises

Simplify each expression.

1. $8(3a - b) + 4(2b - a)$
2. $40s + 18t - 5t + 11s$
3. $\frac{1}{5}(4j + 2k - 6j + 3k)$

4. $10(6g + 3h) + 4(5g - h)$
5. $12\left(\dfrac{a}{3} - \dfrac{b}{4}\right)$
6. $8(2.4r - 3.1s) - 6(1.5r + 2.4s)$

7. $4(20 - 4p) - \frac{3}{4}(4 - 16p)$
8. $5.5j + 8.9k - 4.7k - 10.9j$
9. $1.2(7x - 5) - (10 - 4.3x)$

10. $9(7e - 4f) - 0.6(e + 5f)$
11. $2.5m(12 - 8.5)$
12. $\frac{3}{4}p - \frac{1}{5}r - \frac{3}{5}r - \frac{1}{2}p$

13. $4(10g + 80h) - 20(10h - 5g)$
14. $2(15 + 45c) + \frac{5}{6}(12 + 18c)$

15. $(7 - 2.1x)3 + 2(3.5x - 6)$
16. $\frac{2}{3}(18 - 6n + 12 + 3n)$

17. $14(j - 2) - 3j(4 - 7)$
18. $50(3a - b) - 20(b - 2a)$

4

1-3 Study Guide and Intervention

Solving Equations

Verbal Expressions to Algebraic Expressions The chart suggests some ways to help you translate word expressions into algebraic expressions. Any letter can be used to represent a number that is not known.

Word Expression	Operation
and, plus, sum, increased by, more than	addition
minus, difference, decreased by, less than	subtraction
times, product, of (as in $\frac{1}{2}$ of a number)	multiplication
divided by, quotient	division

Example 1 Write an algebraic expression to represent 18 less than the quotient of a number and 3.

$$\frac{n}{3} - 18$$

Example 2 Write a verbal sentence to represent $6(n - 2) = 14$.

Six times the difference of a number and two is equal to 14.

Exercises

Write an algebraic expression to represent each verbal expression.

1. the sum of six times a number and 25

2. four times the sum of a number and 3

3. 7 less than fifteen times a number

4. the difference of nine times a number and the quotient of 6 and the same number

5. the sum of 100 and four times a number

6. the product of 3 and the sum of 11 and a number

7. four times the square of a number increased by five times the same number

8. 23 more than the product of 7 and a number

Write a verbal sentence to represent each equation.

9. $3n - 35 = 79$

10. $2(n^3 + 3n^2) = 4n$

11. $\dfrac{5n}{n + 3} = n - 8$

1-3 Study Guide and Intervention (continued)

Solving Equations

Properties of Equality You can solve equations by using addition, subtraction, multiplication, or division.

Addition and Subtraction Properties of Equality	For any real numbers a, b, and c, if $a = b$, then $a + c = b + c$ and $a - c = b - c$.
Multiplication and Division Properties of Equality	For any real numbers a, b, and c, if $a = b$, then $a \cdot c = b \cdot c$ and, if c is not zero, $\dfrac{a}{c} = \dfrac{b}{c}$.

Example 1 Solve $100 - 8x = 140$.

$$100 - 8x = 140$$
$$100 - 8x - 100 = 140 - 100$$
$$-8x = 40$$
$$x = -5$$

Example 2 Solve $4x + 5y = 100$ for y.

$$4x + 5y = 100$$
$$4x + 5y - 4x = 100 - 4x$$
$$5y = 100 - 4x$$
$$y = \frac{1}{5}(100 - 4x)$$
$$y = 20 - \frac{4}{5}x$$

Exercises

Solve each equation. Check your solution.

1. $3s = 45$

2. $17 = 9 - a$

3. $5t - 1 = 6t - 5$

4. $\dfrac{2}{3}m = \dfrac{1}{2}$

5. $7 - \dfrac{1}{2}x = 3$

6. $-8 = -2(z + 7)$

7. $0.2b = 10$

8. $3x + 17 = 5x - 13$

9. $5(4 - k) = -10k$

10. $120 - \dfrac{3}{4}y = 60$

11. $\dfrac{5}{2}n = 98 - n$

12. $4.5 + 2p = 8.7$

13. $4n + 20 = 53 - 2n$

14. $100 = 20 - 5r$

15. $2x + 75 = 102 - x$

Solve each equation or formula for the specified variable.

16. $a = 3b - c$, for b

17. $\dfrac{s}{2t} = 10$, for t

18. $h = 12g - 1$, for g

19. $\dfrac{3pq}{r} = 12$, for p

20. $2xy = x + 7$, for x

21. $\dfrac{d}{2} + \dfrac{f}{4} = 6$, for f

22. $3(2j - k) = 108$, for j

23. $3.5s - 42 = 14t$, for s

24. $\dfrac{m}{n} + 5m = 20$, for m

25. $4x - 3y = 10$, for y

1-4 Study Guide and Intervention

Solving Absolute Value Equations

Absolute Value Expressions The **absolute value** of a number is the number of units it is from 0 on a number line. The symbol $|x|$ is used to represent the absolute value of a number x.

Absolute Value	• Words	For any real number a, if a is positive or zero, the absolute value of a is a. If a is negative, the absolute value of a is the opposite of a.
	• Symbols	For any real number a, $|a| = a$, if $a \geq 0$, and $|a| = -a$, if $a < 0$.

Example 1 Evaluate $|-4| - |-2x|$ if $x = 6$.

$$\begin{aligned} |-4| - |-2x| &= |-4| - |-2 \cdot 6| \\ &= |-4| - |-12| \\ &= 4 - 12 \\ &= -8 \end{aligned}$$

Example 2 Evaluate $|2x - 3y|$ if $x = -4$ and $y = 3$.

$$\begin{aligned} |2x - 3y| &= |2(-4) - 3(3)| \\ &= |-8 - 9| \\ &= |-17| \\ &= 17 \end{aligned}$$

Exercises

Evaluate each expression if $w = -4$, $x = 2$, $y = \frac{1}{2}$, and $z = -6$.

1. $|2x - 8|$

2. $|6 + z| - |-7|$

3. $5 + |w + z|$

4. $|x + 5| - |2w|$

5. $|x| - |y| - |z|$

6. $|7 - x| + |3x|$

7. $|w - 4x|$

8. $|wz| - |xy|$

9. $|z| - 3|5yz|$

10. $5|w| + 2|z - 2y|$

11. $|z| - 4|2z + y|$

12. $10 - |xw|$

13. $|6y + z| + |yz|$

14. $3|wx| + \frac{1}{4}|4x + 8y|$

15. $7|yz| - 30$

16. $14 - 2|w - xy|$

17. $|2x - y| + 5y$

18. $|xyz| + |wxz|$

19. $z|z| + x|x|$

20. $12 - |10x - 10y|$

21. $\frac{1}{2}|5z + 8w|$

22. $|yz - 4w| - w$

23. $\frac{3}{4}|wz| + \frac{1}{2}|8y|$

24. $xz - |xz|$

7

1-4 Study Guide and Intervention (continued)

Solving Absolute Value Equations

Absolute Value Equations Use the definition of absolute value to solve equations containing absolute value expressions.

For any real numbers a and b, where $b \geq 0$, if $|a| = b$ then $a = b$ or $a = -b$.

Always check your answers by substituting them into the original equation. Sometimes computed solutions are not actual solutions.

Example Solve $|2x - 3| = 17$. Check your solutions.

Case 1 $a = b$
$$2x - 3 = 17$$
$$2x - 3 + 3 = 17 + 3$$
$$2x = 20$$
$$x = 10$$

CHECK $|2x - 3| = 17$
$$|2(10) - 3| = 17$$
$$|20 - 3| = 17$$
$$|17| = 17$$
$$17 = 17 \checkmark$$

Case 2 $a = -b$
$$2x - 3 = -17$$
$$2x - 3 + 3 = -17 + 3$$
$$2x = -14$$
$$x = -7$$

CHECK $|2(-7) - 3| = 17$
$$|-14 - 3| = 17$$
$$|-17| = 17$$
$$17 = 17 \checkmark$$

There are two solutions, 10 and -7.

Exercises

Solve each equation. Check your solutions.

1. $|x + 15| = 37$

2. $|t - 4| - 5 = 0$

3. $|x - 5| = 45$

4. $|m + 3| = 12 - 2m$

5. $|5b + 9| + 16 = 2$

6. $|15 - 2k| = 45$

7. $5n + 24 = |8 - 3n|$

8. $|8 + 5a| = 14 - a$

9. $\frac{1}{3}|4p - 11| = p + 4$

10. $|3x - 1| = 2x + 11$

11. $\left|\frac{1}{3}x + 3\right| = -1$

12. $40 - 4x = 2|3x - 10|$

13. $5f - |3f + 4| = 20$

14. $|4b + 3| = 15 - 2b$

15. $\frac{1}{2}|6 - 2x| = 3x + 1$

16. $|16 - 3x| = 4x - 12$

8

1-5 Study Guide and Intervention

Solving Inequalities

Solve Inequalities The following properties can be used to solve inequalities.

Addition and Subtraction Properties for Inequalities	Multiplication and Division Properties for Inequalities
For any real numbers a, b, and c: 1. If $a < b$, then $a + c < b + c$ and $a - c < b - c$. 2. If $a > b$, then $a + c > b + c$ and $a - c > b - c$.	For any real numbers a, b, and c, with $c \neq 0$: 1. If c is positive and $a < b$, then $ac < bc$ and $\frac{a}{c} < \frac{b}{c}$. 2. If c is positive and $a > b$, then $ac > bc$ and $\frac{a}{c} > \frac{b}{c}$. 3. If c is negative and $a < b$, then $ac > bc$ and $\frac{a}{c} > \frac{b}{c}$. 4. If c is negative and $a > b$, then $ac < bc$ and $\frac{a}{c} < \frac{b}{c}$.

These properties are also true for $\leq$ and $\geq$.

Example 1 Solve $2x + 4 > 36$. Then graph the solution set on a number line.

$2x + 4 - 4 > 36 - 4$

$\qquad 2x > 32$

$\qquad\quad x > 16$

The solution set is $\{x \mid x > 16\}$.

Example 2 Solve $17 - 3w \geq 35$. Then graph the solution set on a number line.

$17 - 3w \geq 35$

$17 - 3w - 17 \geq 35 - 17$

$\qquad\quad -3w \geq 18$

$\qquad\qquad w \leq -6$

The solution set is $\{w \mid w \leq -6\}$.

Exercises

Solve each inequality. Describe the solution set using set-builder notation. Then graph the solution set on a number line.

1. $7(7a - 9) \leq 84$

2. $3(9z + 4) > 35z - 4$

3. $5(12 - 3n) < 165$

4. $18 - 4k < 2(k + 21)$

5. $4(b - 7) + 6 < 22$

6. $2 + 3(m + 5) \geq 4(m + 3)$

7. $4x - 2 > -7(4x - 2)$

8. $\frac{1}{3}(2y - 3) > y + 2$

9. $2.5d + 15 \leq 75$

1-5 Study Guide and Intervention (continued)

Solving Inequalities

Real-World Problems with Inequalities Many real-world problems involve inequalities. The chart below shows some common phrases that indicate inequalities.

<	>	≤	≥
is less than is fewer than	is greater than is more than	is at most is no more than is less than or equal to	is at least is no less than is greater than or equal to

Example   **SPORTS** The Vikings play 36 games this year. At midseason, they have won 16 games. How many of the remaining games must they win in order to win at least 80% of *all* their games this season?

Let x be the number of remaining games that the Vikings must win. The total number of games they will have won by the end of the season is $16 + x$. They want to win at least 80% of their games. Write an inequality with ≥.

$$16 + x \geq 0.8(36)$$
$$x \geq 0.8(36) - 16$$
$$x \geq 12.8$$

Since they cannot win a fractional part of a game, the Vikings must win at least 13 of the games remaining.

Exercises

1. **PARKING FEES** The city parking lot charges $2.50 for the first hour and $0.25 for each additional hour. If the most you want to pay for parking is $6.50, solve the inequality $2.50 + 0.25(x - 1) \leq 6.50$ to determine for how many hours you can park your car.

PLANNING For Exercises 2 and 3, use the following information.

Ethan is reading a 482-page book for a book report due on Monday. He has already read 80 pages. He wants to figure out how many pages per hour he needs to read in order to finish the book in less than 6 hours.

2. Write an inequality to describe this situation.

3. Solve the inequality and interpret the solution.

BOWLING For Exercises 4 and 5, use the following information.

Four friends plan to spend Friday evening at the bowling alley. Three of the friends need to rent shoes for $3.50 per person. A string (game) of bowling costs $1.50 per person. If the friends pool their $40, how many strings can they afford to bowl?

4. Write an equation to describe this situation.

5. Solve the inequality and interpret the solution.

1-6 Study Guide and Intervention

Solving Compound and Absolute Value Inequalities

Compound Inequalities A compound inequality consists of two inequalities joined by the word *and* or the word *or*. To solve a compound inequality, you must solve each part separately.

And Compound Inequalities	Example: $x > -4$ and $x < 3$ −5 −4 −3 −2 −1 0 1 2 3 4 5	The graph is the intersection of solution sets of two inequalities.
Or Compound Inequalities	Example: $x \leq -3$ or $x > 1$ −5 −4 −3 −2 −1 0 1 2 3 4 5	The graph is the union of solution sets of two inequalities.

Example 1 Solve $-3 \leq 2x + 5 \leq 19$. Graph the solution set on a number line.

$$-3 \leq 2x + 5 \quad \text{and} \quad 2x + 5 \leq 19$$
$$-8 \leq 2x \qquad\qquad\quad 2x \leq 14$$
$$-4 \leq x \qquad\qquad\qquad x \leq 7$$
$$-4 \leq x \leq 7$$

−8 −6 −4 −2 0 2 4 6 8

Example 2 Solve $3y - 2 \geq 7$ or $2y - 1 \leq -9$. Graph the solution set on a number line.

$$3y - 2 \geq 7 \quad \text{or} \quad 2y - 1 \leq -9$$
$$3y \geq 9 \quad \text{or} \qquad 2y \leq -8$$
$$y \geq 3 \quad \text{or} \qquad y \leq -4$$

−8 −6 −4 −2 0 2 4 6 8

Exercises

Solve each inequality. Graph the solution set on a number line.

1. $-10 < 3x + 2 \leq 14$

−8 −6 −4 −2 0 2 4 6 8

2. $3a + 8 < 23$ or $\frac{1}{4}a - 6 > 7$

−10 0 10 20 30 40 50 60 70

3. $18 < 4x - 10 < 50$

3 5 7 9 11 13 15 17 19

4. $5k + 2 < -13$ or $8k - 1 > 19$

−4 −3 −2 −1 0 1 2 3 4

5. $100 \leq 5y - 45 \leq 225$

0 10 20 30 40 50 60 70 80

6. $\frac{2}{3}b - 2 > 10$ or $\frac{3}{4}b + 5 < -4$

−24 −12 0 12 24

7. $22 < 6w - 2 < 82$

0 2 4 6 8 10 12 14 16

8. $4d - 1 > -9$ or $2d + 5 < 11$

−4 −3 −2 −1 0 1 2 3 4

1-6 **Study Guide and Intervention** *(continued)*

Solving Compound and Absolute Value Inequalities

Absolute Value Inequalities Use the definition of absolute value to rewrite an absolute value inequality as a compound inequality.

For all real numbers a and b, $b > 0$, the following statements are true.

1. If $|a| < b$, then $-b < a < b$.
2. If $|a| > b$, then $a > b$ or $a < -b$.

These statements are also true for $\leq$ and $\geq$.

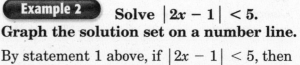

Example 1 Solve $|x + 2| > 4$. Graph the solution set on a number line.

By statement 2 above, if $|x + 2| > 4$, then $x + 2 > 4$ or $x + 2 < -4$. Subtracting 2 from both sides of each inequality gives $x > 2$ or $x < -6$.

$$\overset{\longleftarrow}{\underset{-8\ -6\ -4\ -2\ \ 0\ \ 2\ \ 4\ \ 6\ \ 8}{+\!\!\oplus\!\!+\!\!+\!\!+\!\!+\!\!\oplus\!\!+\!\!+}}\longrightarrow$$

Example 2 Solve $|2x - 1| < 5$. Graph the solution set on a number line.

By statement 1 above, if $|2x - 1| < 5$, then $-5 < 2x - 1 < 5$. Adding 1 to all three parts of the inequality gives $-4 < 2x < 6$. Dividing by 2 gives $-2 < x < 3$.

$$\overset{\longleftarrow}{\underset{-8\ -6\ -4\ -2\ \ 0\ \ 2\ \ 4\ \ 6\ \ 8}{+\!\!+\!\!+\!\!\ominus\!\!+\!\!+\!\!\ominus\!\!+\!\!+}}\longrightarrow$$

Exercises

Solve each inequality. Graph the solution set on a number line.

1. $|3x + 4| < 8$

$$\overset{\longleftarrow}{\underset{-5\ -4\ -3\ -2\ -1\ \ 0\ \ 1\ \ 2\ \ 3}{+\!\!+\!\!+\!\!+\!\!+\!\!+\!\!+\!\!+\!\!+\!\!+}}\longrightarrow$$

2. $|4s| + 1 > 27$

$$\overset{\longleftarrow}{\underset{-8\ -6\ -4\ -2\ \ 0\ \ 2\ \ 4\ \ 6\ \ 8}{+\!\!+\!\!+\!\!+\!\!+\!\!+\!\!+\!\!+\!\!+\!\!+}}\longrightarrow$$

3. $\left|\dfrac{c}{2} - 3\right| \leq 5$

$$\overset{\longleftarrow}{\underset{-8\ -4\ \ 0\ \ 4\ \ 8\ \ 12\ 16\ 20\ 24}{+\!\!+\!\!+\!\!+\!\!+\!\!+\!\!+\!\!+\!\!+}}\longrightarrow$$

4. $|a + 9| \geq 30$

$$\overset{\longleftarrow}{\underset{-40\ \ \ -20\ \ \ \ 0\ \ \ \ 20\ \ \ \ 40}{+\!\!+\!\!+\!\!+\!\!+\!\!+\!\!+\!\!+\!\!+}}\longrightarrow$$

5. $|2f - 11| > 9$

$$\overset{\longleftarrow}{\underset{-4\ -2\ \ 0\ \ 2\ \ 4\ \ 6\ \ 8\ 10\ 12}{+\!\!+\!\!+\!\!+\!\!+\!\!+\!\!+\!\!+\!\!+}}\longrightarrow$$

6. $|5w + 2| < 28$

$$\overset{\longleftarrow}{\underset{-8\ -6\ -4\ -2\ \ 0\ \ 2\ \ 4\ \ 6\ \ 8}{+\!\!+\!\!+\!\!+\!\!+\!\!+\!\!+\!\!+\!\!+\!\!+}}\longrightarrow$$

7. $|10 - 2k| < 2$

$$\overset{\longleftarrow}{\underset{0\ \ 1\ \ 2\ \ 3\ \ 4\ \ 5\ \ 6\ \ 7\ \ 8}{+\!\!+\!\!+\!\!+\!\!+\!\!+\!\!+\!\!+\!\!+}}\longrightarrow$$

8. $\left|\dfrac{x}{2} - 5\right| + 2 > 10$

$$\overset{\longleftarrow}{\underset{-10\ -5\ \ 0\ \ 5\ 10\ 15\ 20\ 25\ 30}{+\!\!+\!\!+\!\!+\!\!+\!\!+\!\!+\!\!+\!\!+\!\!+}}\longrightarrow$$

9. $|4b - 11| < 17$

$$\overset{\longleftarrow}{\underset{-4\ -2\ \ 0\ \ 2\ \ 4\ \ 6\ \ 8\ 10\ 12}{+\!\!+\!\!+\!\!+\!\!+\!\!+\!\!+\!\!+\!\!+}}\longrightarrow$$

10. $|100 - 3m| > 20$

$$\overset{\longleftarrow}{\underset{0\ \ 5\ 10\ 15\ 20\ 25\ 30\ 35\ 40}{+\!\!+\!\!+\!\!+\!\!+\!\!+\!\!+\!\!+\!\!+\!\!+}}\longrightarrow$$

2-1 Study Guide and Intervention

Relations and Functions

Graph Relations A **relation** can be represented as a set of ordered pairs or as an equation; the relation is then the set of all ordered pairs (x, y) that make the equation true. The **domain** of a relation is the set of all first coordinates of the ordered pairs, and the **range** is the set of all second coordinates.

A **function** is a relation in which each element of the domain is paired with exactly one element of the range. You can tell if a relation is a function by graphing, then using the **vertical line test**. If a vertical line intersects the graph at more than one point, the relation is not a function.

Example Graph the equation $y = 2x - 3$ and find the domain and range. Is the equation discrete or continuous? Does the equation represent a function?

Make a table of values to find ordered pairs that satisfy the equation. Then graph the ordered pairs.

The domain and range are both all real numbers. The equation can be graphed by line, so it is continuous. The graph passes the vertical line test, so it is a function.

x	y
−1	−5
0	−3
1	−1
2	1
3	3

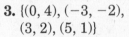

Exercises

Graph each relation or equation and find the domain and range. Next determine if the relation is discrete or continuous. Then determine whether the relation or equation is a function.

1. {(1, 3), (−3, 5), (−2, 5), (2, 3)}

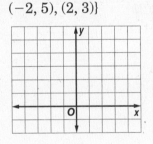

2. {(3, −4), (1, 0), (2, −2), (3, 2)}

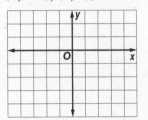

3. {(0, 4), (−3, −2), (3, 2), (5, 1)}

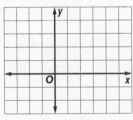

4. $y = x^2 - 1$

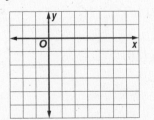

5. $y = x - 4$

6. $y = 3x + 2$

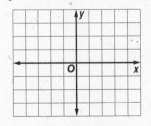

2-1 Study Guide and Intervention (continued)

Relations and Functions

Equations of Functions and Relations Equations that represent functions are often written in **functional notation**. For example, $y = 10 - 8x$ can be written as $f(x) = 10 - 8x$. This notation emphasizes the fact that the values of y, the **dependent variable**, depend on the values of x, the **independent variable**.

To evaluate a function, or find a functional value, means to substitute a given value in the domain into the equation to find the corresponding element in the range.

Example Given the function $f(x) = x^2 + 2x$, find each value.

a. $f(3)$

$$f(x) = x^2 + 2x \qquad \text{Original function}$$
$$f(3) = 3^2 + 2(3) \qquad \text{Substitute.}$$
$$= 15 \qquad \text{Simplify.}$$

b. $f(5a)$

$$f(x) = x^2 + 2x \qquad \text{Original function}$$
$$f(5a) = (5a)^2 + 2(5a) \qquad \text{Substitute.}$$
$$= 25a^2 + 10a \qquad \text{Simplify.}$$

Exercises

Find each value if $f(x) = -2x + 4$.

1. $f(12)$ **2.** $f(6)$ **3.** $f(2b)$

Find each value if $g(x) = x^3 - x$.

4. $g(5)$ **5.** $g(-2)$ **6.** $g(7c)$

Find each value if $f(x) = 2x + \dfrac{2}{x}$ and $g(x) = 0.4x^2 - 1.2$.

7. $f(0.5)$ **8.** $f(-8)$ **9.** $g(3)$

10. $g(-2.5)$ **11.** $f(4a)$ **12.** $g\left(\dfrac{b}{2}\right)$

13. $f\left(\dfrac{1}{3}\right)$ **14.** $g(10)$ **15.** $f(200)$

Let $f(x) = 2x^2 - 1$.

16. Find the values of $f(2)$ and $f(5)$.

17. Compare the values of $f(2) \cdot f(5)$ and $f(2 \cdot 5)$.

2-2 Study Guide and Intervention

Linear Equations

Identify Linear Equations and Functions A **linear equation** has no operations other than addition, subtraction, and multiplication of a variable by a constant. The variables may not be multiplied together or appear in a denominator. A linear equation does not contain variables with exponents other than 1. The graph of a linear equation is a line.

A **linear function** is a function whose ordered pairs satisfy a linear equation. Any linear function can be written in the form $f(x) = mx + b$, where m and b are real numbers.

If an equation is linear, you need only two points that satisfy the equation in order to graph the equation. One way is to find the x-intercept and the y-intercept and connect these two points with a line.

Example 1 Is $f(x) = 0.2 - \dfrac{x}{5}$ a linear function? Explain.

Yes; it is a linear function because it can be written in the form
$f(x) = -\dfrac{1}{5}x + 0.2$.

Example 2 Is $2x + xy - 3y = 0$ a linear function? Explain.

No; it is not a linear function because the variables x and y are multiplied together in the middle term.

Example 3 Find the x-intercept and the y-intercept of the graph of $4x - 5y = 20$. Then graph the equation.

The x-intercept is the value of x when $y = 0$.

$4x - 5y = 20$ Original equation
$4x - 5(0) = 20$ Substitute 0 for y.
$x = 5$ Simplify.

So the x-intercept is 5.
Similarly, the
y-intercept is -4.

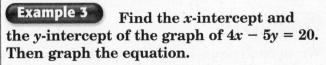

Exercises

State whether each equation or function is linear. Write *yes* or *no*. If no, explain.

1. $6y - x = 7$

2. $9x = \dfrac{18}{y}$

3. $f(x) = 2 - \dfrac{x}{11}$

Find the x-intercept and the y-intercept of the graph of each equation. Then graph the equation.

4. $2x + 7y = 14$

5. $5y - x = 10$

6. $2.5x - 5y + 7.5 = 0$

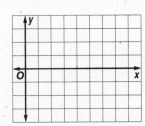

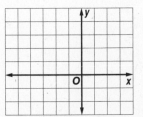

2-2 Study Guide and Intervention (continued)

Linear Equations

Standard Form The **standard form** of a linear equation is $Ax + By = C$, where A, B, and C are integers whose greatest common factor is 1.

Example Write each equation in standard form. Identify A, B, and C.

a. $y = 8x - 5$

$y = 8x - 5$	Original equation
$-8x + y = -5$	Subtract $8x$ from each side.
$8x - y = 5$	Multiply each side by -1.

So $A = 8$, $B = -1$, and $C = 5$.

b. $14x = -7y + 21$

$14x = -7y + 21$	Original equation
$14x + 7y = 21$	Add $7y$ to each side.
$2x + y = 3$	Divide each side by 7.

So $A = 2$, $B = 1$, and $C = 3$.

Exercises

Write each equation in standard form. Identify A, B, and C.

1. $2x = 4y - 1$

2. $5y = 2x + 3$

3. $3x = -5y + 2$

4. $18y = 24x - 9$

5. $\frac{3}{4}y = \frac{2}{3}x + 5$

6. $6y - 8x + 10 = 0$

7. $0.4x + 3y = 10$

8. $x = 4y - 7$

9. $2y = 3x + 6$

10. $\frac{2}{5}x + \frac{1}{3}y - 2 = 0$

11. $4y + 4x + 12 = 0$

12. $3x = -18$

13. $x = \frac{y}{9} + 7$

14. $3y = 9x - 18$

15. $2x = 20 - 8y$

16. $\frac{y}{4} - 3 = 2x$

17. $\left(\frac{5x}{2}\right) = \frac{3}{4}y + 8$

18. $0.25y = 2x - 0.75$

19. $2y - \frac{x}{6} - 4 = 0$

20. $1.6x - 2.4y = 4$

21. $0.2x = 100 - 0.4y$

2-3 Study Guide and Intervention
Slope

Slope

Slope *m* of a Line	For points (x_1, y_1) and (x_2, y_2), where $x_1 \neq x_2$, $m = \dfrac{\text{change in } y}{\text{change in } x} = \dfrac{y_2 - y_1}{x_2 - x_1}$

Example 1 Determine the slope of the line that passes through $(2, -1)$ and $(-4, 5)$.

$m = \dfrac{y_2 - y_1}{x_2 - x_1}$ Slope formula

$= \dfrac{5 - (-1)}{-4 - 2}$ $(x_1, y_1) = (2, -1)$, $(x_2, y_2) = (-4, 5)$

$= \dfrac{6}{-6} = -1$ Simplify.

The slope of the line is -1.

Example 2 Graph the line passing through $(-1, -3)$ with a slope of $\dfrac{4}{5}$.

Graph the ordered pair $(-1, -3)$. Then, according to the slope, go up 4 units and right 5 units. Plot the new point $(4, 1)$. Connect the points and draw the line.

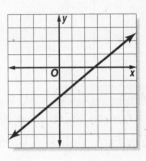

Exercises

Find the slope of the line that passes through each pair of points.

1. $(4, 7)$ and $(6, 13)$

2. $(6, 4)$ and $(3, 4)$

3. $(5, 1)$ and $(7, -3)$

4. $(5, -3)$ and $(-4, 3)$

5. $(5, 10)$ and $(-1, -2)$

6. $(-1, -4)$ and $(-13, 2)$

7. $(7, -2)$ and $(3, 3)$

8. $(-5, 9)$ and $(5, 5)$

9. $(4, -2)$ and $(-4, -8)$

Graph the line passing through the given point with the given slope.

10. slope $= -\dfrac{1}{3}$
passes through $(0, 2)$

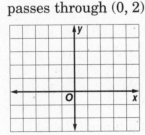

11. slope $= 2$
passes through $(1, 4)$

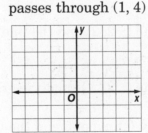

12. slope $= 0$
passes through $(-2, -5)$

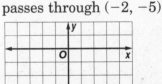

13. slope $= 1$
passes through $(-4, 6)$

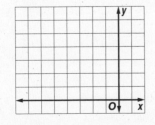

14. slope $= -\dfrac{3}{4}$
passes through $(-3, 0)$

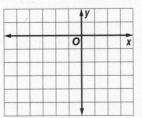

15. slope $= \dfrac{1}{5}$
passes through $(0, 0)$

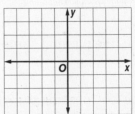

2-3 Study Guide and Intervention *(continued)*

Slope

Parallel and Perpendicular Lines

In a plane, nonvertical lines with the same slope are **parallel**. All vertical lines are parallel.

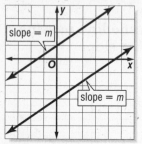

In a plane, two oblique lines are **perpendicular** if and only if the product of their slopes is -1. Any vertical line is perpendicular to any horizontal line.

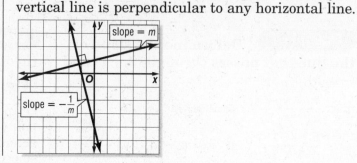

Example Are the line passing through $(2, 6)$ and $(-2, 2)$ and the line passing through $(3, 0)$ and $(0, 4)$ parallel, perpendicular, or neither?

Find the slopes of the two lines.

The slope of the first line is $\dfrac{6 - 2}{2 - (-2)} = 1$.

The slope of the second line is $\dfrac{4 - 0}{0 - 3} = -\dfrac{4}{3}$.

The slopes are not equal and the product of the slopes is not -1, so the lines are neither parallel nor perpendicular.

Exercises

Are the lines parallel, perpendicular, or neither?

1. the line passing through $(4, 3)$ and $(1, -3)$ and the line passing through $(1, 2)$ and $(-1, 3)$

2. the line passing through $(2, 8)$ and $(-2, 2)$ and the line passing through $(0, 9)$ and $(6, 0)$

3. the line passing through $(3, 9)$ and $(-2, -1)$ and the graph of $y = 2x$

4. the line with x-intercept -2 and y-intercept 5 and the line with x-intercept 2 and y-intercept -5

5. the line with x-intercept 1 and y-intercept 3 and the line with x-intercept 3 and y-intercept 1

6. the line passing through $(-2, -3)$ and $(2, 5)$ and the graph of $x + 2y = 10$

7. the line passing through $(-4, -8)$ and $(6, -4)$ and the graph of $2x - 5y = 5$

18

2-4 Study Guide and Intervention

Writing Linear Equations

Forms of Equations

Slope-Intercept Form of a Linear Equation	$y = mx + b$, where m is the slope and b is the y-intercept
Point-Slope Form of a Linear Equation	$y - y_1 = m(x - x_1)$, where (x_1, y_1) are the coordinates of a point on the line and m is the slope of the line

Example 1 Write an equation in slope-intercept form for the line that has slope -2 and passes through the point (3, 7).

Substitute for m, x, and y in the slope-intercept form.

$y = mx + b$ Slope-intercept form
$7 = (-2)(3) + b$ $(x, y) = (3, 7)$, $m = -2$
$7 = -6 + b$ Simplify.
$13 = b$ Add 6 to both sides.

The y-intercept is 13. The equation in slope-intercept form is $y = -2x + 13$.

Example 2 Write an equation in slope-intercept form for the line that has slope $\frac{1}{3}$ and x-intercept 5.

$y = mx + b$ Slope-intercept form
$0 = \left(\frac{1}{3}\right)(5) + b$ $(x, y) = (5, 0)$, $m = \frac{1}{3}$
$0 = \frac{5}{3} + b$ Simplify.
$-\frac{5}{3} = b$ Subtract $\frac{5}{3}$ from both sides.

The y-intercept is $-\frac{5}{3}$. The slope-intercept form is $y = \frac{1}{3}x - \frac{5}{3}$.

Exercises

Write an equation in slope-intercept form for the line that satisfies each set of conditions.

1. slope -2, passes through $(-4, 6)$

2. slope $\frac{3}{2}$, y-intercept 4

3. slope 1, passes through (2, 5)

4. slope $-\frac{13}{5}$, passes through $(5, -7)$

Write an equation in slope-intercept form for each graph.

5.
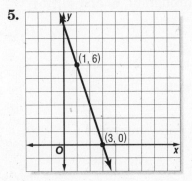

6.
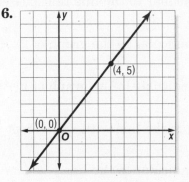

7.
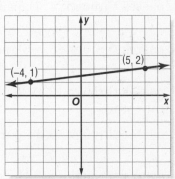

2-4 Study Guide and Intervention (continued)

Writing Linear Equations

Parallel and Perpendicular Lines Use the slope-intercept or point-slope form to find equations of lines that are parallel or perpendicular to a given line. Remember that parallel lines have equal slope. The slopes of two perpendicular lines are negative reciprocals, that is, their product is -1.

Example 1 Write an equation of the line that passes through (8, 2) and is perpendicular to the line whose equation is $y = -\frac{1}{2}x + 3$.

The slope of the given line is $-\frac{1}{2}$. Since the slopes of perpendicular lines are negative reciprocals, the slope of the perpendicular line is 2.

Use the slope and the given point to write the equation.

$y - y_1 = m(x - x_1)$ Point-slope form
$y - 2 = 2(x - 8)$ $(x_1, y_1) = (8, 2)$, $m = 2$
$y - 2 = 2x - 16$ Distributive Prop.
$y = 2x - 14$ Add 2 to each side.

An equation of the line is $y = 2x - 14$.

Example 2 Write an equation of the line that passes through (−1, 5) and is parallel to the graph of $y = 3x + 1$.

The slope of the given line is 3. Since the slopes of parallel lines are equal, the slope of the parallel line is also 3.

Use the slope and the given point to write the equation.

$y - y_1 = m(x - x_1)$ Point-slope form
$y - 5 = 3(x - (-1))$ $(x_1, y_1) = (-1, 5)$, $m = 3$
$y - 5 = 3x + 3$ Distributive Prop.
$y = 3x + 8$ Add 5 to each side.

An equation of the line is $y = 3x + 8$.

Exercises

Write an equation in slope-intercept form for the line that satisfies each set of conditions.

1. passes through (−4, 2), parallel to the line whose equation is $y = \frac{1}{2}x + 5$

2. passes through (3, 1), perpendicular to the graph of $y = -3x + 2$

3. passes through (1, −1), parallel to the line that passes through (4, 1) and (2, −3)

4. passes through (4, 7), perpendicular to the line that passes through (3, 6) and (3, 15)

5. passes through (8, −6), perpendicular to the graph of $2x - y = 4$

6. passes through (2, −2), perpendicular to the graph of $x + 5y = 6$

7. passes through (6, 1), parallel to the line with x-intercept −3 and y-intercept 5

8. passes through (−2, 1), perpendicular to the line $y = 4x - 11$

NAME _____ DATE _____ PERIOD ____

2-5 Study Guide and Intervention
Modeling Real-World Data: Using Scatter Plots

Scatter Plots When a set of data points is graphed as ordered pairs in a coordinate plane, the graph is called a **scatter plot**. A scatter plot can be used to determine if there is a relationship among the data.

Example BASEBALL The table below shows the number of home runs and runs batted in for various baseball players who have won the Most Valuable Player Award since 2002. Make a scatter plot of the data.

Home Runs	Runs Batted In
46	110
34	131
45	90
47	118
45	101
39	126

Source: www.baseball-reference.com

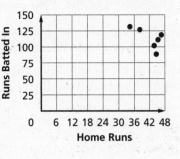

Source: www.baseball-reference.com

Exercises

Make a scatter plot for the data in each table below.

1. **FUEL EFFICIENCY** The table below shows the average fuel efficiency in miles per gallon of vehicles in the U.S. during the years listed.

Year	Fuel Efficiency (mpg)
1970	12.0
1980	13.3
1990	16.4
2000	16.9

Source: U.S. Federal Highway Administration

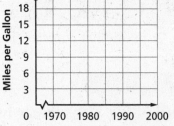

2. **CONGRESS** The table below shows the number of women who served in the United States Congress during the years 1995–2006.

Congressional Session	Number of Women
104	59
105	65
106	67
107	75
108	77
109	83

Source: www.senate.gov

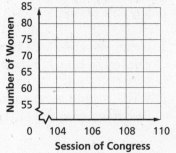

Study Guide and Intervention   21   Glencoe Algebra 2

2-5 Study Guide and Intervention (continued)

Modeling Real-World Data: Using Scatter Plots

Prediction Equations A **line of fit** is a line that closely approximates a set of data graphed in a scatter plot. The equation of a line of fit is called a **prediction equation** because it can be used to predict values not given in the data set.

To find a prediction equation for a set of data, select two points that seem to represent the data well. Then to write the prediction equation, use what you know about writing a linear equation when given two points on the line.

Example   STORAGE COSTS According to a certain prediction equation, the cost of 200 square feet of storage space is $60. The cost of 325 square feet of storage space is $160.

a. Find the slope of the prediction equation. What does it represent?

Since the cost depends upon the square footage, let x represent the amount of storage space in square feet and y represent the cost in dollars. The slope can be found using the

formula $m = \dfrac{y_2 - y_1}{x_2 - x_1}$. So, $m = \dfrac{160 - 60}{325 - 200} = \dfrac{100}{125} = 0.8$

The slope of the prediction equation is 0.8. This means that the price of storage increases 80¢ for each one-square-foot increase in storage space.

b. Find a prediction equation.

Using the slope and one of the points on the line, you can use the point-slope form to find a prediction equation.

$y - y_1 = m(x - x_1)$ Point-slope form
$y - 60 = 0.8(x - 200)$ $(x_1, y_1) = (200, 60)$, $m = 0.8$
$y - 60 = 0.8x - 160$ Distributive Property
$\quad\;\; y = 0.8x - 100$ Add 60 to both sides.

A prediction equation is $y = 0.8x - 100$.

Exercises

SALARIES The table below shows the years of experience for eight technicians at Lewis Techomatic and the hourly rate of pay each technician earns. Use the data for Exercises 1 and 2.

Experience (years)	9	4	3	1	10	6	12	8
Hourly Rate of Pay	$17	$10	$10	$7	$19	$12	$20	$15

1. Draw a scatter plot to show how years of experience are related to hourly rate of pay. Draw a line of fit.

2. Write a prediction equation to show how years of experience (x) are related to hourly rate of pay (y).

Technician Salaries

Copyright © Glencoe/McGraw-Hill, a division of The McGraw-Hill Companies, Inc.

2-6 Study Guide and Intervention

Special Functions

Step Functions, Constant Functions, and the Identity Function The chart below lists some special functions you should be familiar with.

Function	Written as	Graph
Constant	$f(x) = c$	horizontal line
Identity	$f(x) = x$	line through the origin with slope 1
Greatest Integer Function	$f(x) = [\![x]\!]$	one-unit horizontal segments, with right endpoints missing, arranged like steps

The greatest integer function is an example of a **step function**, a function with a graph that consists of horizontal segments.

Example Identify each function as a constant function, the identity function, or a step function.

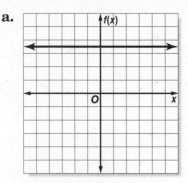

a.

a constant function

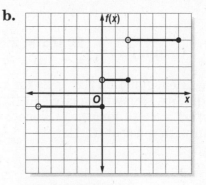

b.

a step function

Exercises

Identify each function as a constant function, the identity function, a greatest integer function, or a step function.

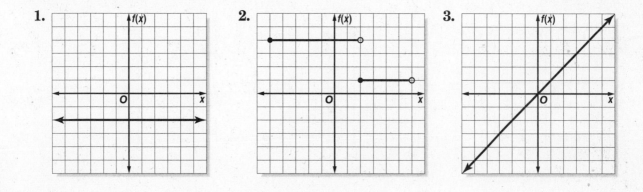

1.

2.

3.

23

2-6 Study Guide and Intervention (continued)

Special Functions

Absolute Value and Piecewise Functions

Function	Written as	Graph
Absolute Value Function	$f(x) = |x|$	two rays that are mirror images of each other and meet at a point, the vertex

The absolute value function can be written as a **piecewise function**. A piecewise function is written using two or more expressions. Its graph is often disjointed.

Example 1 Graph $f(x) = 3|x| - 4$.

Find several ordered pairs. Graph the points and connect them. You would expect the graph to look similar to its parent function, $f(x) = |x|$.

x	$3|x| - 4$
0	−4
1	−1
2	2
−1	−1
−2	2

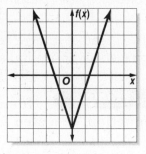

Example 2 Graph $f(x) = \begin{cases} 2x \text{ if } x < 2 \\ x - 1 \text{ if } x \geq 2 \end{cases}$

First, graph the linear function $f(x) = 2x$ for $x < 2$. Since 2 does not satisfy this inequality, stop with a circle at (2, 4). Next, graph the linear function $f(x) = x - 1$ for $x \geq 2$. Since 2 does satisfy this inequality, begin with a dot at (2, 1).

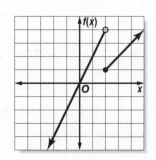

Exercises

Graph each function. Identify the domain and range.

1. $g(x) = \left[\!\left[\dfrac{x}{3}\right]\!\right]$

2. $h(x) = |2x + 1|$

3. $h(x) = \begin{cases} \dfrac{x}{3} \text{ if } x \leq 0 \\ 2x - 6 \text{ if } 0 < x < 2 \\ 1 \text{ if } x \geq 2 \end{cases}$

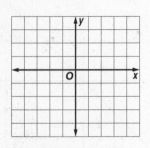

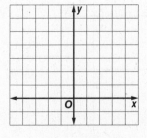

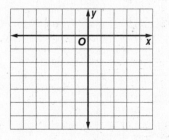

2-7 Study Guide and Intervention

Graphing Inequalities

Graph Linear Inequalities A **linear inequality**, like $y \geq 2x - 1$, resembles a linear equation, but with an inequality sign instead of an equals sign. The graph of the related linear equation separates the coordinate plane into two half-planes. The line is the boundary of each half-plane.

To graph a linear inequality, follow these steps.

1. Graph the boundary; that is, the related linear equation. If the inequality symbol is $\leq$ or $\geq$, the boundary is solid. If the inequality symbol is $<$ or $>$, the boundary is dashed.

2. Choose a point not on the boundary and test it in the inequality. $(0, 0)$ is a good point to choose if the boundary does not pass through the origin.

3. If a true inequality results, shade the half-plane containing your test point. If a false inequality results, shade the other half-plane.

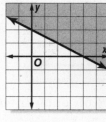

Example Graph $x + 2y \geq 4$.

The boundary is the graph of $x + 2y = 4$.

Use the slope-intercept form, $y = -\dfrac{1}{2}x + 2$, to graph the boundary line.

The boundary line should be solid.

Now test the point $(0, 0)$.

$0 + 2(0) \overset{?}{\geq} 4 \qquad (x, y) = (0, 0)$

$\qquad 0 \geq 4 \qquad$ false

Shade the region that does *not* contain $(0, 0)$.

Exercises

Graph each inequality.

1. $y < 3x + 1$

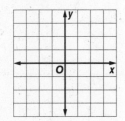

2. $y \geq x - 5$

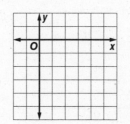

3. $4x + y \leq -1$

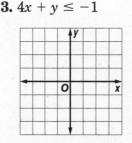

4. $y < \dfrac{x}{2} - 4$

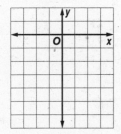

5. $x + y > 6$

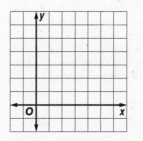

6. $0.5x - 0.25y < 1.5$

2-7 Study Guide and Intervention (continued)

Graphing Inequalities

Graph Absolute Value Inequalities Graphing absolute value inequalities is similar to graphing linear inequalities. The graph of the related absolute value equation is the boundary. This boundary is graphed as a solid line if the inequality is $\leq$ or $\geq$, and dashed if the inequality is $<$ or $>$. Choose a test point not on the boundary to determine which region to shade.

Example Graph $y \leq 3|x - 1|$.

First graph the equation $y = 3|x - 1|$.

Since the inequality is $\leq$, the graph of the boundary is solid.

Test $(0, 0)$.

$0 \overset{?}{\leq} 3|0 - 1|$ $(x, y) = (0, 0)$

$0 \overset{?}{\leq} 3|-1|$ $|-1| = 1$

$0 \leq 3$ true

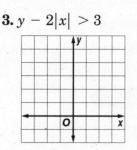

Shade the region that contains $(0, 0)$.

Exercises

Graph each inequality.

1. $y \geq |x| + 1$

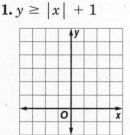

2. $y \leq |2x - 1|$

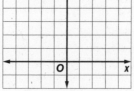

3. $y - 2|x| > 3$

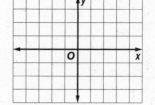

4. $y < -|x| - 3$

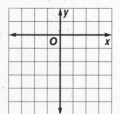

5. $|x| + y \geq 4$

6. $|x + 1| + 2y < 0$

7. $|2 - x| + y > -1$

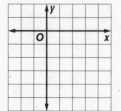

8. $y < 3|x| - 3$

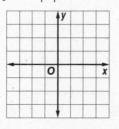

9. $y \leq |1 - x| + 4$

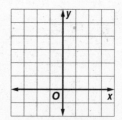

3-1 Study Guide and Intervention

Solving Systems of Equations by Graphing

Graph Systems of Equations A system of equations is a set of two or more equations containing the same variables. You can solve a system of linear equations by graphing the equations on the same coordinate plane. If the lines intersect, the solution is that intersection point.

 Example **Solve the system of equations by graphing.**
$$x - 2y = 4$$
$$x + y = -2$$

Write each equation in slope-intercept form.

$$x - 2y = 4 \quad \rightarrow \quad y = \frac{x}{2} - 2$$

$$x + y = -2 \quad \rightarrow \quad y = -x - 2$$

The graphs appear to intersect at $(0, -2)$.

CHECK Substitute the coordinates into each equation.

$$
\begin{array}{ll}
x - 2y = 4 & x + y = -2 \\
0 - 2(-2) \stackrel{?}{=} 4 & 0 + (-2) \stackrel{?}{=} -2 \\
4 = 4 \ \checkmark & -2 = -2 \ \checkmark
\end{array}
$$

The solution of the system is $(0, -2)$.

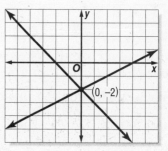

Exercises

Solve each system of equations by graphing.

1. $y = -\dfrac{x}{3} + 1$

$y = \dfrac{x}{2} - 4$

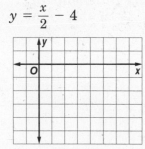

2. $y = 2x - 2$

$y = -x + 4$

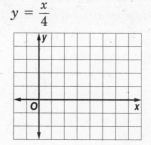

3. $y = -\dfrac{x}{2} + 3$

$y = \dfrac{x}{4}$

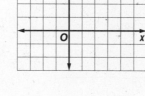

4. $3x - y = 0$

$x - y = -2$

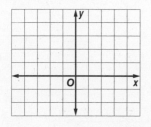

5. $2x + \dfrac{y}{3} = -7$

$\dfrac{x}{2} + y = 1$

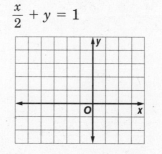

6. $\dfrac{x}{2} - y = 2$

$2x - y = -1$

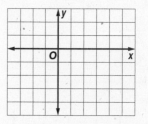

3-1 Study Guide and Intervention (continued)

Solving Systems of Equations by Graphing

Classify Systems of Equations The following chart summarizes the possibilities for graphs of two linear equations in two variables.

Graphs of Equations	Slopes of Lines	Classification of System	Number of Solutions
Lines intersect	Different slopes	Consistent and independent	One
Lines coincide (same line)	Same slope, same y-intercept	Consistent and dependent	Infinitely many
Lines are parallel	Same slope, different y-intercepts	Inconsistent	None

Example Graph the system of equations and describe it as *consistent and independent, consistent and dependent,* or *inconsistent.*

$$x - 3y = 6$$
$$2x - y = -3$$

Write each equation in slope-intercept form.

$$x - 3y = 6 \quad \rightarrow \quad y = \frac{1}{3}x - 2$$
$$2x - y = -3 \quad \rightarrow \quad y = 2x + 3$$

The graphs intersect at $(-3, -3)$. Since there is one solution, the system is consistent and independent.

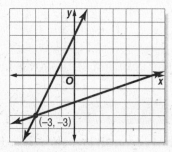

Exercises

Graph the system of equations and describe it as *consistent and independent, consistent and dependent,* or *inconsistent.*

1. $3x + y = -2$
$6x + 2y = 10$

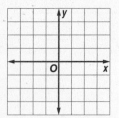

2. $x + 2y = 5$
$3x - 15 = -6y$

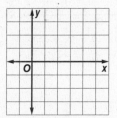

3. $2x - 3y = 0$
$4x - 6y = 3$

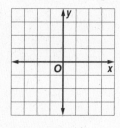

4. $2x - y = 3$
$x + 2y = 4$

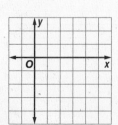

5. $4x + y = -2$
$2x + \frac{y}{2} = -1$

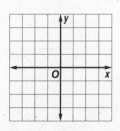

6. $3x - y = 2$
$x + y = 6$

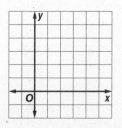

3-2 Study Guide and Intervention

Solving Systems of Equations Algebraically

Substitution To solve a system of linear equations by **substitution**, first solve for one variable in terms of the other in one of the equations. Then substitute this expression into the other equation and simplify.

Example   **Use substitution to solve the system of equations.** $2x - y = 9$
$x + 3y = -6$

Solve the first equation for y in terms of x.

$2x - y = 9$	First equation
$-y = -2x + 9$	Subtract 2x from both sides.
$y = 2x - 9$	Multiply both sides by −1.

Substitute the expression $2x - 9$ for y into the second equation and solve for x.

$x + 3y = -6$	Second equation
$x + 3(2x - 9) = -6$	Substitute 2x − 9 for y.
$x + 6x - 27 = -6$	Distributive Property
$7x - 27 = -6$	Simplify.
$7x = 21$	Add 27 to each side.
$x = 3$	Divide each side by 7.

Now, substitute the value 3 for x in either original equation and solve for y.

$2x - y = 9$	First equation
$2(3) - y = 9$	Replace x with 3.
$6 - y = 9$	Simplify.
$-y = 3$	Subtract 6 from each side.
$y = -3$	Multiply each side by −1.

The solution of the system is $(3, -3)$.

Exercises

Solve each system of linear equations by using substitution.

1. $3x + y = 7$
    $4x + 2y = 16$

2. $2x + y = 5$
    $3x - 3y = 3$

3. $2x + 3y = -3$
    $x + 2y = 2$

4. $2x - y = 7$
    $6x - 3y = 14$

5. $4x - 3y = 4$
    $2x + y = -8$

6. $5x + y = 6$
    $3 - x = 0$

7. $x + 8y = -2$
    $x - 3y = 20$

8. $2x - y = -4$
    $4x + y = 1$

9. $x - y = -2$
    $2x - 3y = 2$

10. $x - 4y = 4$
    $2x + 12y = 13$

11. $x + 3y = 2$
    $4x + 12y = 8$

12. $2x + 2y = 4$
    $x - 2y = 0$

3-2 Study Guide and Intervention (continued)
Solving Systems of Equations Algebraically

Elimination To solve a system of linear equations by **elimination**, add or subtract the equations to eliminate one of the variables. You may first need to multiply one or both of the equations by a constant so that one of the variables has the same (or opposite) coefficient in one equation as it has in the other.

Example 1 Use the elimination method to solve the system of equations.

$$2x - 4y = -26$$
$$3x - y = -24$$

Multiply the second equation by 4. Then subtract the equations to eliminate the y variable.

$$2x - 4y = -26$$
$$3x - y = -24 \quad \text{Multiply by 4.}$$

$$
\begin{array}{r}
2x - 4y = -26 \\
12x - 4y = -96 \\
\hline
-10x \quad\quad = 70 \\
x \quad\quad = -7
\end{array}
$$

Replace x with -7 and solve for y.

$$2x - 4y = -26$$
$$2(-7) - 4y = -26$$
$$-14 - 4y = -26$$
$$-4y = -12$$
$$y = 3$$

The solution is $(-7, 3)$.

Example 2 Use the elimination method to solve the system of equations.

$$3x - 2y = 4$$
$$5x + 3y = -25$$

Multiply the first equation by 3 and the second equation by 2. Then add the equations to eliminate the y variable.

$$3x - 2y = 4 \quad \text{Multiply by 3.}$$
$$5x + 3y = -25 \quad \text{Multiply by 2.}$$

$$
\begin{array}{r}
9x - 6y = 12 \\
10x + 6y = -50 \\
\hline
19x \quad\quad = -38 \\
x \quad\quad = -2
\end{array}
$$

Replace x with -2 and solve for y.

$$3x - 2y = 4$$
$$3(-2) - 2y = 4$$
$$-6 - 2y = 4$$
$$-2y = 10$$
$$y = -5$$

The solution is $(-2, -5)$.

Exercises

Solve each system of equations by using elimination.

1. $2x - y = 7$
$3x + y = 8$

2. $x - 2y = 4$
$-x + 6y = 12$

3. $3x + 4y = -10$
$x - 4y = 2$

4. $3x - y = 12$
$5x + 2y = 20$

5. $4x - y = 6$
$2x - \dfrac{y}{2} = 4$

6. $5x + 2y = 12$
$-6x - 2y = -14$

7. $2x + y = 8$
$3x + \dfrac{3}{2}y = 12$

8. $7x + 2y = -1$
$4x - 3y = -13$

9. $3x + 8y = -6$
$x - y = 9$

10. $5x + 4y = 12$
$7x - 6y = 40$

11. $-4x + y = -12$
$4x + 2y = 6$

12. $5m + 2n = -8$
$4m + 3n = 2$

3-3 Study Guide and Intervention

Solving Systems of Inequalities by Graphing

Graph Systems of Inequalities To solve a system of inequalities, graph the inequalities in the same coordinate plane. The solution set is represented by the intersection of the graphs.

Example Solve the system of inequalities by graphing.

$y \leq 2x - 1$ and $y > \dfrac{x}{3} + 2$

The solution of $y \leq 2x - 1$ is Regions 1 and 2.

The solution of $y > \dfrac{x}{3} + 2$ is Regions 1 and 3.

The intersection of these regions is Region 1, which is the solution set of the system of inequalities.

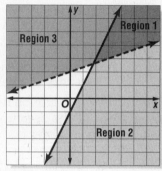

Exercises

Solve each system of inequalities by graphing.

1. $x - y \leq 2$
   $x + 2y \geq 1$

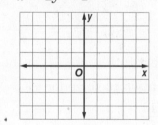

2. $3x - 2y \leq -1$
   $x + 4y \geq -12$

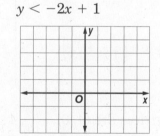

3. $|y| \leq 1$
   $x > 2$

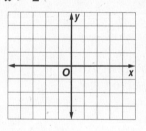

4. $y \geq \dfrac{x}{2} - 3$

   $y < 2x$

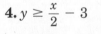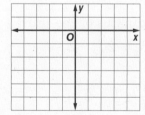

5. $y < \dfrac{x}{3} + 2$

   $y < -2x + 1$

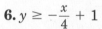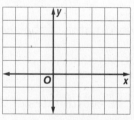

6. $y \geq -\dfrac{x}{4} + 1$

   $y < 3x - 1$

7. $x + y \geq 4$
   $2x - y > 2$

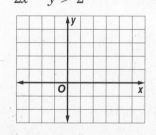

8. $x + 3y < 3$
   $x - 2y \geq 4$

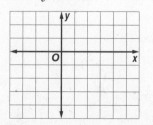

9. $x - 2y > 6$
   $x + 4y < -4$

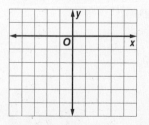

3-3 Study Guide and Intervention (continued)

Solving Systems of Inequalities by Graphing

Find Vertices of a Polygonal Region Sometimes the graph of a system of inequalities forms a bounded region. You can find the vertices of the region by a combination of the methods used earlier in this chapter: graphing, substitution, and/or elimination.

Example Find the coordinates of the vertices of the figure formed by $5x + 4y < 20$, $y < 2x + 3$, and $x - 3y < 4$.

Graph the boundary of each inequality. The intersections of the boundary lines are the vertices of a triangle.

The vertex $(4, 0)$ can be determined from the graph. To find the coordinates of the second and third vertices, solve the two systems of equations

$$y = 2x + 3 \qquad y = 2x + 3$$
$$5x + 4y = 20 \quad \text{and} \quad x - 3y = 4$$

For the first system of equations, rewrite the first equation in standard form as $2x - y = -3$. Then multiply that equation by 4 and add to the second equation.

$$2x - y = -3 \quad \text{Multiply by 4.}$$
$$5x + 4y = 20$$

$$\begin{aligned} 8x - 4y &= -12 \\ (+) \quad 5x + 4y &= 20 \\ \hline 13x &= 8 \\ x &= \frac{8}{13} \end{aligned}$$

Then substitute $x = \frac{8}{13}$ in one of the original equations and solve for y.

$$2\left(\frac{8}{13}\right) - y = -3$$

$$\frac{16}{13} - y = -3$$

$$y = \frac{55}{13}$$

The coordinates of the second vertex are $\left(\frac{8}{13}, 4\frac{3}{13}\right)$.

For the second system of equations, use substitution.

Substitute $2x + 3$ for y in the second equation to get

$$x - 3(2x + 3) = 4$$
$$x - 6x - 9 = 4$$
$$-5x = 13$$
$$x = -\frac{13}{5}$$

Then substitute $x = -\frac{13}{5}$ in the first equation to solve for y.

$$y = 2\left(-\frac{13}{5}\right) + 3$$

$$y = -\frac{26}{5} + 3$$

$$y = -\frac{11}{5}$$

The coordinates of the third vertex are $\left(-2\frac{3}{5}, -2\frac{1}{5}\right)$.

Thus, the coordinates of the three vertices are $(4, 0)$, $\left(\frac{8}{13}, 4\frac{3}{13}\right)$, and $\left(-2\frac{3}{5}, -2\frac{1}{5}\right)$.

Exercises

Find the coordinates of the vertices of the figure formed by each system of inequalities.

1. $y \leq -3x + 7$

  $y < \frac{1}{2}x$

  $y > -2$

2. $x > -3$

  $y < -\frac{1}{3}x + 3$

  $y > x - 1$

3. $y < -\frac{1}{2}x + 3$

  $y > \frac{1}{2}x + 1$

  $y < 3x + 10$

32

3-4 Study Guide and Intervention

Linear Programming

Maximum and Minimum Values When a system of linear inequalities produces a bounded polygonal region, the *maximum* or *minimum* value of a related function will occur at a vertex of the region.

Example Graph the system of inequalities. Name the coordinates of the vertices of the feasible region. Find the maximum and minimum values of the function $f(x, y) = 3x + 2y$ for this polygonal region.

$y \le 4$

$y \le -x + 6$

$y \ge \frac{1}{2}x - \frac{3}{2}$

$y \le 6x + 4$

First find the vertices of the bounded region. Graph the inequalities.

The polygon formed is a quadrilateral with vertices at $(0, 4)$, $(2, 4)$, $(5, 1)$, and $(-1, -2)$. Use the table to find the maximum and minimum values of $f(x, y) = 3x + 2y$.

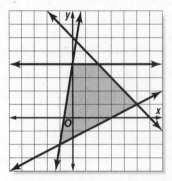

(x, y)	$3x + 2y$	$f(x, y)$
$(0, 4)$	$3(0) + 2(4)$	8
$(2, 4)$	$3(2) + 2(4)$	14
$(5, 1)$	$3(5) + 2(1)$	17
$(-1, -2)$	$3(-1) + 2(-2)$	-7

The maximum value is 17 at $(5, 1)$. The minimum value is -7 at $(-1, -2)$.

Exercises

Graph each system of inequalities. Name the coordinates of the vertices of the feasible region. Find the maximum and minimum values of the given function for this region.

1. $y \ge 2$
$1 \le x \le 5$
$y \le x + 3$
$f(x, y) = 3x - 2y$

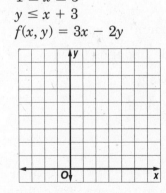

2. $y \ge -2$
$y \ge 2x - 4$
$x - 2y \ge -1$
$f(x, y) = 4x - y$

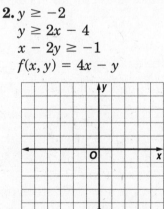

3. $x + y \ge 2$
$4y \le x + 8$
$y \ge 2x - 5$
$f(x, y) = 4x + 3y$

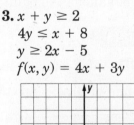
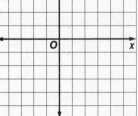

3-4 Study Guide and Intervention (continued)

Linear Programming

Real-World Problems When solving **linear programming** problems, use the following procedure.

1. Define variables.
2. Write a system of inequalities.
3. Graph the system of inequalities.
4. Find the coordinates of the vertices of the feasible region.
5. Write an expression to be maximized or minimized.
6. Substitute the coordinates of the vertices in the expression.
7. Select the greatest or least result to answer the problem.

Example A painter has exactly 32 units of yellow dye and 54 units of green dye. He plans to mix as many gallons as possible of color A and color B. Each gallon of color A requires 4 units of yellow dye and 1 unit of green dye. Each gallon of color B requires 1 unit of yellow dye and 6 units of green dye. Find the maximum number of gallons he can mix.

Step 1 Define the variables.

x = the number of gallons of color A made

y = the number of gallons of color B made

Step 2 Write a system of inequalities.

Since the number of gallons made cannot be negative, $x \geq 0$ and $y \geq 0$.

There are 32 units of yellow dye; each gallon of color A requires 4 units, and each gallon of color B requires 1 unit.

So $4x + y \leq 32$.

Similarly for the green dye, $x + 6y \leq 54$.

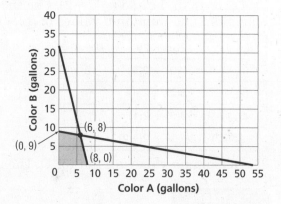

Steps 3 and 4 Graph the system of inequalities and find the coordinates of the vertices of the feasible region. The vertices of the feasible region are (0, 0), (0, 9), (6, 8), and (8, 0).

Steps 5–7 Find the maximum number of gallons, $x + y$, that he can make. The maximum number of gallons the painter can make is 14, 6 gallons of color A and 8 gallons of color B.

Exercises

1. **FOOD** A delicatessen has 12 pounds of plain sausage and 10 pounds of spicy sausage. A pound of Bratwurst A contains $\frac{3}{4}$ pound of plain sausage and $\frac{1}{4}$ pound of spicy sausage. A pound of Bratwurst B contains $\frac{1}{2}$ pound of each sausage.

   Find the maximum number of pounds of bratwurst that can be made.

2. **MANUFACTURING** Machine A can produce 30 steering wheels per hour at a cost of $8 per hour. Machine B can produce 40 steering wheels per hour at a cost of $12 per hour. The company can use either machine by itself or both machines at the same time. What is the minimum number of hours needed to produce 380 steering wheels if the cost must be no more than $108?

3-5 Study Guide and Intervention

Solving Systems of Equations in Three Variables

Systems in Three Variables Use the methods used for solving systems of linear equations in two variables to solve systems of equations in three variables. A system of three equations in three variables can have a unique solution, infinitely many solutions, or no solution. A solution is an **ordered triple**.

Example Solve this system of equations.
$$3x + y - z = -6$$
$$2x - y + 2z = 8$$
$$4x + y - 3z = -21$$

Step 1 Use elimination to make a system of two equations in two variables.

$3x + y - z = -6$ First equation $2x - y + 2z = 8$ Second equation
$(+) 2x - y + 2z = 8$ Second equation $(+) 4x + y - 3z = -21$ Third equation
$\overline{5x + \quad z = 2}$ Add to eliminate y. $\overline{6x \quad - z = -13}$ Add to eliminate y.

Step 2 Solve the system of two equations.

$5x + z = 2$
$(+) 6x - z = -13$
$\overline{11x \quad = -11}$ Add to eliminate z.
$x = -1$ Divide both sides by 11.

Substitute -1 for x in one of the equations with two variables and solve for z.

$5x + z = 2$ Equation with two variables
$5(-1) + z = 2$ Replace x with -1.
$-5 + z = 2$ Multiply.
$z = 7$ Add 5 to both sides.

The result so far is $x = -1$ and $z = 7$.

Step 3 Substitute -1 for x and 7 for z in one of the original equations with three variables.

$3x + y - z = -6$ Original equation with three variables
$3(-1) + y - 7 = -6$ Replace x with -1 and z with 7.
$-3 + y - 7 = -6$ Multiply.
$y = 4$ Simplify.

The solution is $(-1, 4, 7)$.

Exercises

Solve each system of equations.

1. $2x + 3y - z = 0$
$x - 2y - 4z = 14$
$3x + y - 8z = 17$

2. $2x - y + 4z = 11$
$x + 2y - 6z = -11$
$3x - 2y - 10z = 11$

3. $x - 2y + z = 8$
$2x + y - z = 0$
$3x - 6y + 3z = 24$

4. $3x - y - z = 5$
$3x + 2y - z = 11$
$6x - 3y + 2z = -12$

5. $2x - 4y - z = 10$
$4x - 8y - 2z = 16$
$3x + y + z = 12$

6. $x - 6y + 4z = 2$
$2x + 4y - 8z = 16$
$x - 2y = 5$

3-5 Study Guide and Intervention (continued)

Solving Systems of Equations in Three Variables

Real-World Problems

Example The Laredo Sports Shop sold 10 balls, 3 bats, and 2 bases for $99 on Monday. On Tuesday they sold 4 balls, 8 bats, and 2 bases for $78. On Wednesday they sold 2 balls, 3 bats, and 1 base for $33.60. What are the prices of 1 ball, 1 bat, and 1 base?

First define the variables.

x = price of 1 ball
y = price of 1 bat
z = price of 1 base

Translate the information in the problem into three equations.

$10x + 3y + 2z = 99$
$4x + 8y + 2z = 78$
$2x + 3y + z = 33.60$

Subtract the second equation from the first equation to eliminate z.

$$\begin{array}{r} 10x + 3y + 2z = 99 \\ (-)\ \ 4x + 8y + 2z = 78 \\ \hline 6x - 5y \quad\quad = 21 \end{array}$$

Multiply the third equation by 2 and subtract from the second equation.

$$\begin{array}{r} 4x + 8y + 2z = 78 \\ (-)\ 4x + 6y + 2z = 67.20 \\ \hline 2y \quad\quad = 10.80 \\ y \quad = 5.40 \end{array}$$

Substitute 5.40 for y in the equation $6x - 5y = 21$.

$6x - 5(5.40) = 21$
$6x = 48$
$x = 8$

Substitute 8 for x and 5.40 for y in one of the original equations to solve for z.

$10x + 3y + 2z = 99$
$10(8) + 3(5.40) + 2z = 99$
$80 + 16.20 + 2z = 99$
$2z = 2.80$
$z = 1.40$

So a ball costs $8, a bat $5.40, and a base $1.40.

Exercises

1. **FITNESS TRAINING** Carly is training for a triathlon. In her training routine each week, she runs 7 times as far as she swims, and she bikes 3 times as far as she runs. One week she trained a total of 232 miles. How far did she run that week?

2. **ENTERTAINMENT** At the arcade, Ryan, Sara, and Tim played video racing games, pinball, and air hockey. Ryan spent $6 for 6 racing games, 2 pinball games, and 1 game of air hockey. Sara spent $12 for 3 racing games, 4 pinball games, and 5 games of air hockey. Tim spent $12.25 for 2 racing games, 7 pinball games, and 4 games of air hockey. How much did each of the games cost?

3. **FOOD** A natural food store makes its own brand of trail mix out of dried apples, raisins, and peanuts. One pound of the mixture costs $3.18. It contains twice as much peanuts by weight as apples. One pound of dried apples costs $4.48, a pound of raisins $2.40, and a pound of peanuts $3.44. How many ounces of each ingredient are contained in 1 pound of the trail mix?

4-1 Study Guide and Intervention

Introduction to Matrices

Organize Data

Matrix	a rectangular array of variables or constants in horizontal rows and vertical columns, usually enclosed in brackets.

A matrix can be described by its **dimensions**. A matrix with m rows and n columns is an $m \times n$ matrix.

Example 1 Owls' eggs incubate for 30 days and their fledgling period is also 30 days. Swifts' eggs incubate for 20 days and their fledgling period is 44 days. Pigeon eggs incubate for 15 days, and their fledgling period is 17 days. Eggs of the king penguin incubate for 53 days, and the fledgling time for a king penguin is 360 days. Write a 2 × 4 matrix to organize this information. **Source:** *The Cambridge Factfinder*

$$
\begin{array}{c}
\quad\;\; \text{Owl}\;\; \text{Swift}\;\; \text{Pigeon}\;\; \text{King Penguin} \\
\begin{array}{c} \text{Incubation} \\ \text{Fledgling} \end{array}
\begin{bmatrix} 30 & 20 & 15 & 53 \\ 30 & 44 & 17 & 360 \end{bmatrix}
\end{array}
$$

Example 2 What are the dimensions of matrix A if $A = \begin{bmatrix} 13 & 10 & -3 & 45 \\ 2 & 8 & 15 & 80 \end{bmatrix}$?

Since matrix A has 2 rows and 4 columns, the dimensions of A are 2×4.

Exercises

State the dimensions of each matrix.

1. $\begin{bmatrix} 15 & 5 & 27 & -4 \\ 23 & 6 & 0 & 5 \\ 14 & 70 & 24 & -3 \\ 63 & 3 & 42 & 90 \end{bmatrix}$

2. $[16 \quad 12 \quad 0]$

3. $\begin{bmatrix} 71 & 44 \\ 39 & 27 \\ 45 & 16 \\ 92 & 53 \\ 78 & 65 \end{bmatrix}$

4. A travel agent provides for potential travelers the normal high temperatures for the months of January, April, July, and October for various cities. In Boston these figures are 36°, 56°, 82°, and 63°. In Dallas they are 54°, 76°, 97°, and 79°. In Los Angeles they are 68°, 72°, 84°, and 79°. In Seattle they are 46°, 58°, 74°, and 60°, and in St. Louis they are 38°, 67°, 89°, and 69°. Organize this information in a 4 × 5 matrix. **Source:** *The New York Times Almanac*

4-1 Study Guide and Intervention (continued)

Introduction to Matrices

Equations Involving Matrices

Equal Matrices	Two matrices are equal if they have the same dimensions and each element of one matrix is equal to the corresponding element of the other matrix.

You can use the definition of equal matrices to solve matrix equations.

Example Solve $\begin{bmatrix} 4x \\ y \end{bmatrix} = \begin{bmatrix} -2y + 2 \\ x - 8 \end{bmatrix}$ for x and y.

Since the matrices are equal, the corresponding elements are equal. When you write the sentences to show the equality, two linear equations are formed.

$4x = -2y + 2$
$y = x - 8$

This system can be solved using substitution.

$4x = -2y + 2$ First equation
$4x = -2(x - 8) + 2$ Substitute $x - 8$ for y.
$4x = -2x + 16 + 2$ Distributive Property
$6x = 18$ Add $2x$ to each side.
$x = 3$ Divide each side by 6.

To find the value of y, substitute 3 for x in either equation.

$y = x - 8$ Second equation
$y = 3 - 8$ Substitute 3 for x.
$y = -5$ Subtract.

The solution is $(3, -5)$.

Exercises

Solve each equation.

1. $[5x \; 4y] = [20 \; 20]$

2. $\begin{bmatrix} 3x \\ y \end{bmatrix} = \begin{bmatrix} 28 + 4y \\ -3x - 2 \end{bmatrix}$

3. $\begin{bmatrix} -2y \\ x \end{bmatrix} = \begin{bmatrix} 4 - 5x \\ y + 5 \end{bmatrix}$

4. $\begin{bmatrix} x - 2y \\ 3x - 4y \end{bmatrix} = \begin{bmatrix} -1 \\ 22 \end{bmatrix}$

5. $\begin{bmatrix} 2x + 3y \\ x - 2y \end{bmatrix} = \begin{bmatrix} 3 \\ 12 \end{bmatrix}$

6. $\begin{bmatrix} 5x + 3y \\ 2x - y \end{bmatrix} = \begin{bmatrix} -1 \\ -18 \end{bmatrix}$

7. $\begin{bmatrix} 8x - y & 16x \\ 12 & y - 4x \end{bmatrix} = \begin{bmatrix} 18 & 20 \\ 12 & -13 \end{bmatrix}$

8. $\begin{bmatrix} 8x - 6y \\ 12x + 4y \end{bmatrix} = \begin{bmatrix} -3 \\ -11 \end{bmatrix}$

9. $\begin{bmatrix} \frac{x}{3} + \frac{y}{7} \\ \frac{x}{2} + 2y \end{bmatrix} = \begin{bmatrix} 9 \\ 51 \end{bmatrix}$

10. $\begin{bmatrix} 3x + 1.5 \\ 2y - 2.4 \end{bmatrix} = \begin{bmatrix} 7.5 \\ 8.0 \end{bmatrix}$

11. $\begin{bmatrix} 2x + 3y \\ -4x + 0.5y \end{bmatrix} = \begin{bmatrix} 17 \\ -8 \end{bmatrix}$

12. $\begin{bmatrix} x - y \\ x + y \end{bmatrix} = \begin{bmatrix} 0 \\ -25 \end{bmatrix}$

4-2 Study Guide and Intervention

Operations with Matrices

Add and Subtract Matrices

Addition of Matrices	$\begin{bmatrix} a & b & c \\ d & e & f \\ g & h & i \end{bmatrix} + \begin{bmatrix} j & k & l \\ m & n & o \\ p & q & r \end{bmatrix} = \begin{bmatrix} a+j & b+k & c+l \\ d+m & e+n & f+o \\ g+p & h+q & i+r \end{bmatrix}$
Subtraction of Matrices	$\begin{bmatrix} a & b & c \\ d & e & f \\ g & h & i \end{bmatrix} - \begin{bmatrix} j & k & l \\ m & n & o \\ p & q & r \end{bmatrix} = \begin{bmatrix} a-j & b-k & c-l \\ d-m & e-n & f-o \\ g-p & h-q & i-r \end{bmatrix}$

Example 1 Find $A + B$ if $A = \begin{bmatrix} 6 & -7 \\ 2 & -12 \end{bmatrix}$ and $B = \begin{bmatrix} 4 & 2 \\ -5 & -6 \end{bmatrix}$.

$A + B = \begin{bmatrix} 6 & -7 \\ 2 & -12 \end{bmatrix} + \begin{bmatrix} 4 & 2 \\ -5 & -6 \end{bmatrix}$

$= \begin{bmatrix} 6+4 & -7+2 \\ 2+(-5) & -12+(-6) \end{bmatrix}$

$= \begin{bmatrix} 10 & -5 \\ -3 & -18 \end{bmatrix}$

Example 2 Find $A - B$ if $A = \begin{bmatrix} -2 & 8 \\ 3 & -4 \\ 10 & 7 \end{bmatrix}$ and $B = \begin{bmatrix} 4 & -3 \\ -2 & 1 \\ -6 & 8 \end{bmatrix}$.

$A - B = \begin{bmatrix} -2 & 8 \\ 3 & -4 \\ 10 & 7 \end{bmatrix} - \begin{bmatrix} 4 & -3 \\ -2 & 1 \\ -6 & 8 \end{bmatrix}$

$= \begin{bmatrix} -2-4 & 8-(-3) \\ 3-(-2) & -4-1 \\ 10-(-6) & 7-8 \end{bmatrix} = \begin{bmatrix} -6 & 11 \\ 5 & -5 \\ 16 & -1 \end{bmatrix}$

Exercises

Perform the indicated operations. If the matrix does not exist, write **impossible**.

1. $\begin{bmatrix} 8 & 7 \\ -10 & -6 \end{bmatrix} - \begin{bmatrix} -4 & 3 \\ 2 & -12 \end{bmatrix}$

2. $\begin{bmatrix} 6 & -5 & 9 \\ -3 & 4 & 5 \end{bmatrix} + \begin{bmatrix} -4 & 3 & 2 \\ 6 & 9 & -4 \end{bmatrix}$

3. $\begin{bmatrix} 6 \\ -3 \\ 2 \end{bmatrix} + \begin{bmatrix} -6 & 3 & -2 \end{bmatrix}$

4. $\begin{bmatrix} 5 & -2 \\ -4 & 6 \\ 7 & 9 \end{bmatrix} + \begin{bmatrix} -11 & 6 \\ 2 & -5 \\ 4 & -7 \end{bmatrix}$

5. $\begin{bmatrix} 8 & 0 & -6 \\ 4 & 5 & -11 \\ -7 & 3 & 4 \end{bmatrix} - \begin{bmatrix} -2 & 1 & 7 \\ 3 & -4 & 3 \\ -8 & 5 & 6 \end{bmatrix}$

6. $\begin{bmatrix} \frac{3}{4} & \frac{2}{5} \\ -\frac{1}{2} & \frac{4}{3} \end{bmatrix} - \begin{bmatrix} \frac{1}{2} & \frac{2}{3} \\ \frac{2}{3} & -\frac{1}{2} \end{bmatrix}$

4-2 Study Guide and Intervention (continued)

Operations with Matrices

Scalar Multiplication You can multiply an $m \times n$ matrix by a scalar k.

Scalar Multiplication	$k \begin{bmatrix} a & b & c \\ d & e & f \end{bmatrix} = \begin{bmatrix} ka & kb & kc \\ kd & ke & kf \end{bmatrix}$

Example If $A = \begin{bmatrix} 4 & 0 \\ -6 & 3 \end{bmatrix}$ and $B = \begin{bmatrix} -1 & 5 \\ 7 & 8 \end{bmatrix}$, find $3B - 2A$.

$3B - 2A = 3\begin{bmatrix} -1 & 5 \\ 7 & 8 \end{bmatrix} - 2\begin{bmatrix} 4 & 0 \\ -6 & 3 \end{bmatrix}$ Substitution

$= \begin{bmatrix} 3(-1) & 3(5) \\ 3(7) & 3(8) \end{bmatrix} - \begin{bmatrix} 2(4) & 2(0) \\ 2(-6) & 2(3) \end{bmatrix}$ Multiply.

$= \begin{bmatrix} -3 & 15 \\ 21 & 24 \end{bmatrix} - \begin{bmatrix} 8 & 0 \\ -12 & 6 \end{bmatrix}$ Simplify.

$= \begin{bmatrix} -3 - 8 & 15 - 0 \\ 21 - (-12) & 24 - 6 \end{bmatrix}$ Subtract.

$= \begin{bmatrix} -11 & 15 \\ 33 & 18 \end{bmatrix}$ Simplify.

Exercises

Perform the indicated matrix operations. If the matrix does not exist, write *impossible.*

1. $6\begin{bmatrix} 2 & -5 & 3 \\ 0 & 7 & -1 \\ -4 & 6 & 9 \end{bmatrix}$

2. $-\dfrac{1}{3}\begin{bmatrix} 6 & 15 & 9 \\ 51 & -33 & 24 \\ -18 & 3 & 45 \end{bmatrix}$

3. $0.2\begin{bmatrix} 25 & -10 & -45 \\ 5 & 55 & -30 \\ 60 & 35 & -95 \end{bmatrix}$

4. $3\begin{bmatrix} -4 & 5 \\ 2 & 3 \end{bmatrix} - 2\begin{bmatrix} -1 & 2 \\ -3 & 5 \end{bmatrix}$

5. $-2\begin{bmatrix} 3 & -1 \\ 0 & 7 \end{bmatrix} + 4\begin{bmatrix} -2 & 0 \\ 2 & 5 \end{bmatrix}$

6. $2\begin{bmatrix} 6 & -10 \\ -5 & 8 \end{bmatrix} + 5\begin{bmatrix} 2 & 1 \\ 4 & 3 \end{bmatrix}$

7. $4\begin{bmatrix} 1 & -2 & 5 \\ -3 & 4 & 1 \end{bmatrix} - 2\begin{bmatrix} 4 & 3 & -4 \\ 2 & -5 & -1 \end{bmatrix}$

8. $8\begin{bmatrix} 2 & 1 \\ 3 & -1 \\ -2 & 4 \end{bmatrix} + 3\begin{bmatrix} 4 & 0 \\ -2 & 3 \\ 3 & -4 \end{bmatrix}$

9. $\dfrac{1}{4}\left(\begin{bmatrix} 9 & 1 \\ -7 & 0 \end{bmatrix} + \begin{bmatrix} 3 & -5 \\ 1 & 7 \end{bmatrix} \right)$

4-3　Study Guide and Intervention

Multiplying Matrices

Multiply Matrices You can multiply two matrices if and only if the number of columns in the first matrix is equal to the number of rows in the second matrix.

Multiplication of Matrices	$\begin{bmatrix} a_1 & b_1 \\ a_2 & b_2 \end{bmatrix} \cdot \begin{bmatrix} x_1 & y_1 \\ x_2 & y_2 \end{bmatrix} = \begin{bmatrix} a_1x_1 + b_1x_2 & a_1y_1 + b_1y_2 \\ a_2x_1 + b_2x_2 & a_2y_1 + b_2y_2 \end{bmatrix}$

Example Find AB if $A = \begin{bmatrix} -4 & 3 \\ 2 & -2 \\ 1 & 7 \end{bmatrix}$ and $B = \begin{bmatrix} 5 & -2 \\ -1 & 3 \end{bmatrix}$.

$AB = \begin{bmatrix} -4 & 3 \\ 2 & -2 \\ 1 & 7 \end{bmatrix} \cdot \begin{bmatrix} 5 & -2 \\ -1 & 3 \end{bmatrix}$　　　Substitution

$= \begin{bmatrix} -4(5) + 3(-1) & -4(-2) + 3(3) \\ 2(5) + (-2)(-1) & 2(-2) + (-2)(3) \\ 1(5) + 7(-1) & 1(-2) + 7(3) \end{bmatrix}$　　Multiply columns by rows.

$= \begin{bmatrix} -23 & 17 \\ 12 & -10 \\ -2 & 19 \end{bmatrix}$　　　Simplify.

Exercises

Find each product, if possible.

1. $\begin{bmatrix} 4 & 1 \\ -2 & 3 \end{bmatrix} \cdot \begin{bmatrix} 3 & 0 \\ 0 & 3 \end{bmatrix}$

2. $\begin{bmatrix} -1 & 0 \\ 3 & 7 \end{bmatrix} \cdot \begin{bmatrix} 3 & 2 \\ -1 & 4 \end{bmatrix}$

3. $\begin{bmatrix} 3 & -1 \\ 2 & 4 \end{bmatrix} \cdot \begin{bmatrix} 3 & -1 \\ 2 & 4 \end{bmatrix}$

4. $\begin{bmatrix} -3 & 1 \\ 5 & -2 \end{bmatrix} \cdot \begin{bmatrix} 4 & 0 & -2 \\ -3 & 1 & 1 \end{bmatrix}$

5. $\begin{bmatrix} 3 & -2 \\ 0 & 4 \\ -5 & 1 \end{bmatrix} \cdot \begin{bmatrix} 1 & 2 \\ 2 & 1 \end{bmatrix}$

6. $\begin{bmatrix} 5 & -2 \\ 2 & -3 \end{bmatrix} \cdot \begin{bmatrix} 4 & -1 \\ -2 & 5 \end{bmatrix}$

7. $\begin{bmatrix} 6 & 10 \\ -4 & 3 \\ -2 & 7 \end{bmatrix} \cdot \begin{bmatrix} 0 & 4 & -3 \end{bmatrix}$

8. $\begin{bmatrix} 7 & -2 \\ 5 & -4 \end{bmatrix} \cdot \begin{bmatrix} 1 & -3 \\ -2 & 0 \end{bmatrix}$

9. $\begin{bmatrix} 2 & 0 & -3 \\ 1 & 4 & -2 \\ -1 & 3 & 1 \end{bmatrix} \cdot \begin{bmatrix} 2 & -2 \\ 3 & 1 \\ -2 & 4 \end{bmatrix}$

4-3 Study Guide and Intervention (continued)

Multiplying Matrices

Multiplicative Properties The Commutative Property of Multiplication does *not* hold for matrices.

Properties of Matrix Multiplication	For any matrices A, B, and C for which the matrix product is defined, and any scalar c, the following properties are true.
Associative Property of Matrix Multiplication	$(AB)C = A(BC)$
Associative Property of Scalar Multiplication	$c(AB) = (cA)B = A(cB)$
Left Distributive Property	$C(A + B) = CA + CB$
Right Distributive Property	$(A + B)C = AC + BC$

Example Use $A = \begin{bmatrix} 4 & -3 \\ 2 & 1 \end{bmatrix}$, $B = \begin{bmatrix} 2 & 0 \\ 5 & -3 \end{bmatrix}$, and $C = \begin{bmatrix} 1 & -2 \\ 6 & 3 \end{bmatrix}$ to find each product.

a. $(A + B)C$

$$(A + B)C = \left(\begin{bmatrix} 4 & -3 \\ 2 & 1 \end{bmatrix} + \begin{bmatrix} 2 & 0 \\ 5 & -3 \end{bmatrix} \right) \cdot \begin{bmatrix} 1 & -2 \\ 6 & 3 \end{bmatrix}$$

$$= \begin{bmatrix} 6 & -3 \\ 7 & -2 \end{bmatrix} \cdot \begin{bmatrix} 1 & -2 \\ 6 & 3 \end{bmatrix}$$

$$= \begin{bmatrix} 6(1) + (-3)(6) & 6(-2) + (-3)(3) \\ 7(1) + (-2)(6) & 7(-2) + (-2)(3) \end{bmatrix}$$

$$= \begin{bmatrix} -12 & -21 \\ -5 & -20 \end{bmatrix}$$

b. $AC + BC$

$$AC + BC = \begin{bmatrix} 4 & -3 \\ 2 & 1 \end{bmatrix} \cdot \begin{bmatrix} 1 & -2 \\ 6 & 3 \end{bmatrix} + \begin{bmatrix} 2 & 0 \\ 5 & -3 \end{bmatrix} \cdot \begin{bmatrix} 1 & -2 \\ 6 & 3 \end{bmatrix}$$

$$= \begin{bmatrix} 4(1) + (-3)(6) & 4(-2) + (-3)(3) \\ 2(1) + 1(6) & 2(-2) + 1(3) \end{bmatrix} + \begin{bmatrix} 2(1) + 0(6) & 2(-2) + 0(3) \\ 5(1) + (-3)(6) & 5(-2) + (-3)(3) \end{bmatrix}$$

$$= \begin{bmatrix} -14 & -17 \\ 8 & -1 \end{bmatrix} + \begin{bmatrix} 2 & -4 \\ -13 & -19 \end{bmatrix} = \begin{bmatrix} -12 & -21 \\ -5 & -20 \end{bmatrix}$$

Note that although the results in the example illustrate the Right Distributive Property, they do not prove it.

Exercises

Use $A = \begin{bmatrix} 3 & 2 \\ 5 & -2 \end{bmatrix}$, $B = \begin{bmatrix} 6 & 4 \\ 2 & 1 \end{bmatrix}$, $C = \begin{bmatrix} -\frac{1}{2} & -2 \\ 1 & -3 \end{bmatrix}$, and scalar $c = -4$ to determine whether each of the following equations is true for the given matrices.

1. $c(AB) = (cA)B$

2. $AB = BA$

3. $BC = CB$

4. $(AB)C = A(BC)$

5. $C(A + B) = AC + BC$

6. $c(A + B) = cA + cB$

42

4-4 Study Guide and Intervention

Transformations with Matrices

Translations and Dilations Matrices that represent coordinates of points on a plane are useful in describing transformations.

Translation	a transformation that moves a figure from one location to another on the coordinate plane

You can use matrix addition and a translation matrix to find the coordinates of the translated figure.

Dilation	a transformation in which a figure is enlarged or reduced

You can use scalar multiplication to perform dilations.

Example **Find the coordinates of the vertices of the image of $\triangle ABC$ with vertices $A(-5, 4)$, $B(-1, 5)$, and $C(-3, -1)$ if it is moved 6 units to the right and 4 units down. Then graph $\triangle ABC$ and its image $\triangle A'B'C'$.**

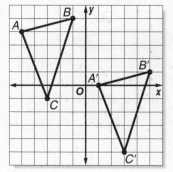

Write the vertex matrix for $\triangle ABC$. $\begin{bmatrix} -5 & -1 & -3 \\ 4 & 5 & -1 \end{bmatrix}$

Add the translation matrix $\begin{bmatrix} 6 & 6 & 6 \\ -4 & -4 & -4 \end{bmatrix}$ to the vertex matrix of $\triangle ABC$.

$$\begin{bmatrix} -5 & -1 & -3 \\ 4 & 5 & -1 \end{bmatrix} + \begin{bmatrix} 6 & 6 & 6 \\ -4 & -4 & -4 \end{bmatrix} = \begin{bmatrix} 1 & 5 & 3 \\ 0 & 1 & -5 \end{bmatrix}$$

The coordinates of the vertices of $\triangle A'B'C'$ are $A'(1, 0)$, $B'(5, 1)$, and $C'(3, -5)$.

Exercises

For Exercises 1 and 2 use the following information. Quadrilateral $QUAD$ with vertices $Q(-1, -3)$, $U(0, 0)$, $A(5, -1)$, and $D(2, -5)$ is translated 3 units to the left and 2 units up.

1. Write the translation matrix.

2. Find the coordinates of the vertices of $Q'U'A'D'$.

For Exercises 3–5, use the following information. The vertices of $\triangle ABC$ are $A(4, -2)$, $B(2, 8)$, and $C(8, 2)$. The triangle is dilated so that its perimeter is one-fourth the original perimeter.

3. Write the coordinates of the vertices of $\triangle ABC$ in a vertex matrix.

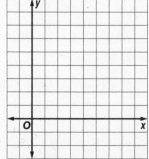

4. Find the coordinates of the vertices of image $\triangle A'B'C'$.

5. Graph the preimage and the image.

4-4 Study Guide and Intervention (continued)

Transformations with Matrices

Reflections and Rotations

Reflection Matrices	For a reflection over the:	x-axis	y-axis	line y = x
	multiply the vertex matrix on the left by:	$\begin{bmatrix} 1 & 0 \\ 0 & -1 \end{bmatrix}$	$\begin{bmatrix} -1 & 0 \\ 0 & 1 \end{bmatrix}$	$\begin{bmatrix} 0 & 1 \\ 1 & 0 \end{bmatrix}$
Rotation Matrices	For a counterclockwise rotation about the origin of:	90°	180°	270°
	multiply the vertex matrix on the left by:	$\begin{bmatrix} 0 & -1 \\ 1 & 0 \end{bmatrix}$	$\begin{bmatrix} -1 & 0 \\ 0 & -1 \end{bmatrix}$	$\begin{bmatrix} 0 & 1 \\ -1 & 0 \end{bmatrix}$

Example Find the coordinates of the vertices of the image of $\triangle ABC$ with $A(3, 5)$, $B(-2, 4)$, and $C(1, -1)$ after a reflection over the line $y = x$.

Write the ordered pairs as a vertex matrix. Then multiply the vertex matrix by the reflection matrix for $y = x$.

$$\begin{bmatrix} 0 & 1 \\ 1 & 0 \end{bmatrix} \cdot \begin{bmatrix} 3 & -2 & 1 \\ 5 & 4 & -1 \end{bmatrix} = \begin{bmatrix} 5 & 4 & -1 \\ 3 & -2 & 1 \end{bmatrix}$$

The coordinates of the vertices of $A'B'C'$ are $A'(5, 3)$, $B'(4, -2)$, and $C'(-1, 1)$.

Exercises

1. The coordinates of the vertices of quadrilateral $ABCD$ are $A(-2, 1)$, $B(-1, 3)$, $C(2, 2)$, and $D(2, -1)$. What are the coordinates of the vertices of the image $A'B'C'D'$ after a reflection over the y-axis?

2. Triangle DEF with vertices $D(-2, 5)$, $E(1, 4)$, and $F(0, -1)$ is rotated 90° counterclockwise about the origin.

   a. Write the coordinates of the triangle in a vertex matrix.

   b. Write the rotation matrix for this situation.

   c. Find the coordinates of the vertices of $\triangle D'E'F'$.

   d. Graph $\triangle DEF$ and $\triangle D'E'F'$.

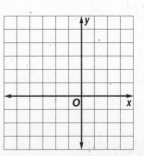

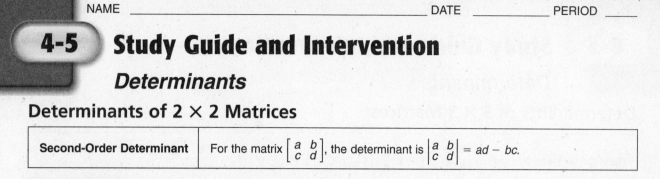

4-5 Study Guide and Intervention

Determinants

Determinants of 2 × 2 Matrices

Second-Order Determinant	For the matrix $\begin{bmatrix} a & b \\ c & d \end{bmatrix}$, the determinant is $\begin{vmatrix} a & b \\ c & d \end{vmatrix} = ad - bc$.

Example Find the value of each determinant.

a. $\begin{vmatrix} 6 & 3 \\ -8 & 5 \end{vmatrix}$

$\begin{vmatrix} 6 & 3 \\ -8 & 5 \end{vmatrix} = 6(5) - 3(-8)$

$= 30 - (-24) \text{ or } 54$

b. $\begin{vmatrix} 11 & -5 \\ 9 & 3 \end{vmatrix}$

$\begin{vmatrix} 11 & -5 \\ 9 & 3 \end{vmatrix} = 11(-3) - (-5)(9)$

$= -33 - (-45) \text{ or } 78$

Exercises

Find the value of each determinant.

1. $\begin{vmatrix} 6 & -2 \\ 5 & 7 \end{vmatrix}$

2. $\begin{vmatrix} -8 & 3 \\ -2 & 1 \end{vmatrix}$

3. $\begin{vmatrix} 3 & 9 \\ 4 & 6 \end{vmatrix}$

4. $\begin{vmatrix} 5 & 12 \\ -7 & -4 \end{vmatrix}$

5. $\begin{vmatrix} -6 & -3 \\ -4 & -1 \end{vmatrix}$

6. $\begin{vmatrix} 4 & 7 \\ 5 & 9 \end{vmatrix}$

7. $\begin{vmatrix} 14 & 8 \\ 9 & -3 \end{vmatrix}$

8. $\begin{vmatrix} 15 & 12 \\ 23 & 28 \end{vmatrix}$

9. $\begin{vmatrix} -8 & 35 \\ 5 & 20 \end{vmatrix}$

10. $\begin{vmatrix} 10 & 16 \\ 22 & 40 \end{vmatrix}$

11. $\begin{vmatrix} 24 & -8 \\ 7 & -3 \end{vmatrix}$

12. $\begin{vmatrix} 13 & 62 \\ -4 & 19 \end{vmatrix}$

13. $\begin{vmatrix} 0.2 & 8 \\ -1.5 & 15 \end{vmatrix}$

14. $\begin{vmatrix} 8.6 & 0.5 \\ 14 & 5 \end{vmatrix}$

15. $\begin{vmatrix} 20 & 110 \\ 0.1 & 1.4 \end{vmatrix}$

16. $\begin{vmatrix} 4.8 & 2.1 \\ 3.4 & 5.3 \end{vmatrix}$

17. $\begin{vmatrix} \frac{2}{3} & -\frac{1}{2} \\ \frac{1}{6} & \frac{1}{5} \end{vmatrix}$

18. $\begin{vmatrix} 6.8 & 15 \\ -0.2 & 5 \end{vmatrix}$

4-5 Study Guide and Intervention *(continued)*

Determinants

Determinants of 3 × 3 Matrices

Third-Order Determinants	$\begin{vmatrix} a & b & c \\ d & e & f \\ g & h & i \end{vmatrix} = a\begin{vmatrix} e & f \\ h & i \end{vmatrix} - b\begin{vmatrix} d & f \\ g & i \end{vmatrix} + c\begin{vmatrix} d & e \\ g & h \end{vmatrix}$

Area of a Triangle	The area of a triangle having vertices (a, b), (c, d) and (e, f) is $|A|$, where $A = \frac{1}{2}\begin{vmatrix} a & b & 1 \\ c & d & 1 \\ e & f & 1 \end{vmatrix}$.

Example Evaluate $\begin{vmatrix} 4 & 5 & -2 \\ 1 & 3 & 0 \\ 2 & -3 & 6 \end{vmatrix}$.

$\begin{vmatrix} 4 & 5 & -2 \\ 1 & 3 & 0 \\ 2 & -3 & 6 \end{vmatrix} = 4\begin{vmatrix} 3 & 0 \\ -3 & 6 \end{vmatrix} - 5\begin{vmatrix} 1 & 0 \\ 2 & 6 \end{vmatrix} - 2\begin{vmatrix} 1 & 3 \\ 2 & -3 \end{vmatrix}$ Third-order determinant

$= 4(18 - 0) - 5(6 - 0) - 2(-3 - 6)$ Evaluate 2 × 2 determinants.

$= 4(18) - 5(6) - 2(-9)$ Simplify.

$= 72 - 30 + 18$ Multiply.

$= 60$ Simplify.

Exercises

Evaluate each determinant.

1. $\begin{vmatrix} 3 & -2 & -2 \\ 0 & 4 & 1 \\ -1 & 5 & -3 \end{vmatrix}$

2. $\begin{vmatrix} 4 & 1 & 0 \\ -2 & 3 & 1 \\ 2 & -2 & 5 \end{vmatrix}$

3. $\begin{vmatrix} 6 & 1 & 4 \\ -2 & 3 & 0 \\ -1 & 3 & 2 \end{vmatrix}$

4. $\begin{vmatrix} 5 & -2 & 2 \\ 3 & 0 & -2 \\ 2 & 4 & -3 \end{vmatrix}$

5. $\begin{vmatrix} 6 & 1 & -4 \\ 3 & 2 & 1 \\ -2 & 2 & -1 \end{vmatrix}$

6. $\begin{vmatrix} 5 & -4 & 1 \\ 2 & 3 & -2 \\ -1 & 6 & -3 \end{vmatrix}$

7. Find the area of a triangle with vertices $X(2, -3)$, $Y(7, 4)$, and $Z(-5, 5)$.

4-6 Study Guide and Intervention
Cramer's Rule

Systems of Two Linear Equations Determinants provide a way for solving systems of equations.

Cramer's Rule for Two-Variable Systems	The solution of the linear system of equations $ax + by = e$ $cx + dy = f$ is (x, y) where $x = \dfrac{\begin{vmatrix} e & b \\ f & d \end{vmatrix}}{\begin{vmatrix} a & b \\ c & d \end{vmatrix}}$, $y = \dfrac{\begin{vmatrix} a & e \\ c & f \end{vmatrix}}{\begin{vmatrix} a & b \\ c & d \end{vmatrix}}$, and $\begin{vmatrix} a & b \\ c & d \end{vmatrix} \neq 0$.

Example Use Cramer's Rule to solve the system of equations. $5x - 10y = 8$
$$10x + 25y = -2$$

$x = \dfrac{\begin{vmatrix} e & b \\ f & d \end{vmatrix}}{\begin{vmatrix} a & b \\ c & d \end{vmatrix}}$ Cramer's Rule $y = \dfrac{\begin{vmatrix} a & e \\ c & f \end{vmatrix}}{\begin{vmatrix} a & b \\ c & d \end{vmatrix}}$

$= \dfrac{\begin{vmatrix} 8 & -10 \\ -2 & 25 \end{vmatrix}}{\begin{vmatrix} 5 & -10 \\ 10 & 25 \end{vmatrix}}$ $a = 5, b = -10, c = 10, d = 25, e = 8, f = -2$ $= \dfrac{\begin{vmatrix} 5 & 8 \\ 10 & -2 \end{vmatrix}}{\begin{vmatrix} 5 & -10 \\ 10 & 25 \end{vmatrix}}$

$= \dfrac{8(25) - (-2)(-10)}{5(25) - (-10)(10)}$ Evaluate each determinant. $= \dfrac{5(-2) - 8(10)}{5(25) - (-10)(10)}$

$= \dfrac{180}{225}$ or $\dfrac{4}{5}$ Simplify. $= -\dfrac{90}{225}$ or $-\dfrac{2}{5}$

The solution is $\left(\dfrac{4}{5}, -\dfrac{2}{5} \right)$.

Exercises

Use Cramer's Rule to solve each system of equations.

1. $3x - 2y = 7$
$2x + 7y = 38$

2. $x - 4y = 17$
$3x - y = 29$

3. $2x - y = -2$
$4x - y = 4$

4. $2x - y = 1$
$5x + 2y = -29$

5. $4x + 2y = 1$
$5x - 4y = 24$

6. $6x - 3y = -3$
$2x + y = 21$

7. $2x + 7y = 16$
$x - 2y = 30$

8. $2x - 3y = -2$
$3x - 4y = 9$

9. $\dfrac{x}{3} + \dfrac{y}{5} = 2$
$\dfrac{x}{4} - \dfrac{y}{6} = -8$

10. $6x - 9y = -1$
$3x + 18y = 12$

11. $3x - 12y = -14$
$9x + 6y = -7$

12. $8x + 2y = \dfrac{3}{7}$
$5x - 4y = -\dfrac{27}{7}$

4-6 Study Guide and Intervention (continued)

Cramer's Rule

Systems of Three Linear Equations

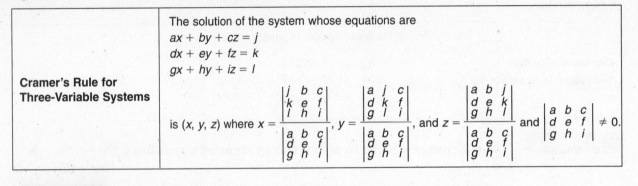

Cramer's Rule for Three-Variable Systems	The solution of the system whose equations are $ax + by + cz = j$ $dx + ey + fz = k$ $gx + hy + iz = l$ is (x, y, z) where $x = \dfrac{\begin{vmatrix} j & b & c \\ k & e & f \\ l & h & i \end{vmatrix}}{\begin{vmatrix} a & b & c \\ d & e & f \\ g & h & i \end{vmatrix}}$, $y = \dfrac{\begin{vmatrix} a & j & c \\ d & k & f \\ g & l & i \end{vmatrix}}{\begin{vmatrix} a & b & c \\ d & e & f \\ g & h & i \end{vmatrix}}$, and $z = \dfrac{\begin{vmatrix} a & b & j \\ d & e & k \\ g & h & l \end{vmatrix}}{\begin{vmatrix} a & b & c \\ d & e & f \\ g & h & i \end{vmatrix}}$ and $\begin{vmatrix} a & b & c \\ d & e & f \\ g & h & i \end{vmatrix} \neq 0$.

Example   **Use Cramer's rule to solve the system of equations.**

$6x + 4y + z = 5$
$2x + 3y - 2z = -2$
$8x - 2y + 2z = 10$

Use the coefficients and constants from the equations to form the determinants. Then evaluate each determinant.

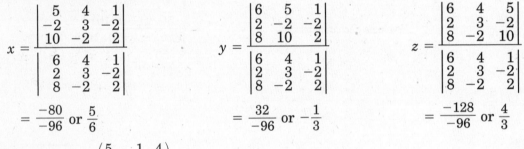

$$x = \frac{\begin{vmatrix} 5 & 4 & 1 \\ -2 & 3 & -2 \\ 10 & -2 & 2 \end{vmatrix}}{\begin{vmatrix} 6 & 4 & 1 \\ 2 & 3 & -2 \\ 8 & -2 & 2 \end{vmatrix}}$$

$$= \frac{-80}{-96} \text{ or } \frac{5}{6}$$

$$y = \frac{\begin{vmatrix} 6 & 5 & 1 \\ 2 & -2 & -2 \\ 8 & 10 & 2 \end{vmatrix}}{\begin{vmatrix} 6 & 4 & 1 \\ 2 & 3 & -2 \\ 8 & -2 & 2 \end{vmatrix}}$$

$$= \frac{32}{-96} \text{ or } -\frac{1}{3}$$

$$z = \frac{\begin{vmatrix} 6 & 4 & 5 \\ 2 & 3 & -2 \\ 8 & -2 & 10 \end{vmatrix}}{\begin{vmatrix} 6 & 4 & 1 \\ 2 & 3 & -2 \\ 8 & -2 & 2 \end{vmatrix}}$$

$$= \frac{-128}{-96} \text{ or } \frac{4}{3}$$

The solution is $\left(\dfrac{5}{6}, -\dfrac{1}{3}, \dfrac{4}{3}\right)$.

Exercises

Use Cramer's rule to solve each system of equations.

1. $x - 2y + 3z = 6$
    $2x - y - z = -3$
    $x + y + z = 6$

2. $3x + y - 2z = -2$
    $4x - 2y - 5z = 7$
    $x + y + z = 1$

3. $x - 3y + z = 1$
    $2x + 2y - z = -8$
    $4x + 7y + 2z = 11$

4. $2x - y + 3z = -5$
    $x + y - 5z = 21$
    $3x - 2y - 4z = 6$

5. $3x + y - 4z = 7$
    $2x - y + 5z = -24$
    $10x + 3y - 2z = -2$

6. $2x - y + 4z = 9$
    $3x - 2y - 5z = -13$
    $x + y - 7z = 0$

4-7 Study Guide and Intervention

Identity and Inverse Matrices

Identity and Inverse Matrices The identity matrix for matrix multiplication is a square matrix with 1s for every element of the main diagonal and zeros elsewhere.

Identity Matrix for Multiplication	If A is an $n \times n$ matrix and I is the identity matrix, then $A \cdot I = A$ and $I \cdot A = A$.

If an $n \times n$ matrix A has an inverse A^{-1}, then $A \cdot A^{-1} = A^{-1} \cdot A = I$.

Example Determine whether $X = \begin{bmatrix} 7 & 4 \\ 10 & 6 \end{bmatrix}$ and $Y = \begin{bmatrix} 3 & -2 \\ -5 & \frac{7}{2} \end{bmatrix}$ are inverse matrices.

Find $X \cdot Y$.

$$X \cdot Y = \begin{bmatrix} 7 & 4 \\ 10 & 6 \end{bmatrix} \cdot \begin{bmatrix} 3 & -2 \\ -5 & \frac{7}{2} \end{bmatrix}$$

$$= \begin{bmatrix} 21 - 20 & -14 + 14 \\ 30 - 30 & -20 + 21 \end{bmatrix} \text{ or } \begin{bmatrix} 1 & 0 \\ 0 & 1 \end{bmatrix}$$

Find $Y \cdot X$.

$$Y \cdot X = \begin{bmatrix} 3 & -2 \\ -5 & \frac{7}{2} \end{bmatrix} \cdot \begin{bmatrix} 7 & 4 \\ 10 & 6 \end{bmatrix}$$

$$= \begin{bmatrix} 21 - 20 & 12 - 12 \\ -35 + 35 & -20 + 21 \end{bmatrix} \text{ or } \begin{bmatrix} 1 & 0 \\ 0 & 1 \end{bmatrix}$$

Since $X \cdot Y = Y \cdot X = I$, X and Y are inverse matrices.

Exercises

Determine whether each pair of matrices are inverses.

1. $\begin{bmatrix} 4 & 5 \\ 3 & 4 \end{bmatrix}$ and $\begin{bmatrix} 4 & -5 \\ -3 & 4 \end{bmatrix}$

2. $\begin{bmatrix} 3 & 2 \\ 5 & 4 \end{bmatrix}$ and $\begin{bmatrix} 2 & -1 \\ -\frac{5}{2} & \frac{3}{2} \end{bmatrix}$

3. $\begin{bmatrix} 2 & 3 \\ 5 & -1 \end{bmatrix}$ and $\begin{bmatrix} 2 & 3 \\ -1 & -2 \end{bmatrix}$

4. $\begin{bmatrix} 8 & 11 \\ 3 & 4 \end{bmatrix}$ and $\begin{bmatrix} -4 & 11 \\ 3 & -8 \end{bmatrix}$

5. $\begin{bmatrix} 4 & -1 \\ 5 & 3 \end{bmatrix}$ and $\begin{bmatrix} 1 & 2 \\ 3 & 8 \end{bmatrix}$

6. $\begin{bmatrix} 5 & 2 \\ 11 & 4 \end{bmatrix}$ and $\begin{bmatrix} -2 & 1 \\ \frac{11}{2} & -\frac{5}{2} \end{bmatrix}$

7. $\begin{bmatrix} 4 & 2 \\ 6 & -2 \end{bmatrix}$ and $\begin{bmatrix} -\frac{1}{5} & -\frac{1}{10} \\ \frac{3}{10} & \frac{1}{10} \end{bmatrix}$

8. $\begin{bmatrix} 5 & 8 \\ 4 & 6 \end{bmatrix}$ and $\begin{bmatrix} -3 & 4 \\ 2 & -\frac{5}{2} \end{bmatrix}$

9. $\begin{bmatrix} 3 & 7 \\ 2 & 4 \end{bmatrix}$ and $\begin{bmatrix} \frac{7}{2} & -\frac{3}{2} \\ 1 & -2 \end{bmatrix}$

10. $\begin{bmatrix} 3 & 2 \\ 4 & -6 \end{bmatrix}$ and $\begin{bmatrix} 3 & 2 \\ -4 & -3 \end{bmatrix}$

11. $\begin{bmatrix} 7 & 2 \\ 17 & 5 \end{bmatrix}$ and $\begin{bmatrix} 5 & -2 \\ -17 & 7 \end{bmatrix}$

12. $\begin{bmatrix} 4 & 3 \\ 7 & 5 \end{bmatrix}$ and $\begin{bmatrix} -5 & 3 \\ 7 & -4 \end{bmatrix}$

4-7 Study Guide and Intervention (continued)

Identity and Inverse Matrices

Find Inverse Matrices

Inverse of a 2 × 2 Matrix	The inverse of a matrix $A = \begin{bmatrix} a & b \\ c & d \end{bmatrix}$ is $A^{-1} = \frac{1}{ad - bc} \begin{bmatrix} d & -b \\ -c & a \end{bmatrix}$, where $ad - bc \neq 0$.

If $ad - bc = 0$, the matrix does not have an inverse.

Example Find the inverse of $N = \begin{bmatrix} 7 & 2 \\ 2 & 1 \end{bmatrix}$.

First find the value of the determinant.

$\begin{vmatrix} 7 & 2 \\ 2 & 1 \end{vmatrix} = 7 - 4 = 3$

Since the determinant does not equal 0, N^{-1} exists.

$$N^{-1} = \frac{1}{ad - bc} \begin{bmatrix} d & -b \\ -c & a \end{bmatrix} = \frac{1}{3} \begin{bmatrix} 1 & -2 \\ -2 & 7 \end{bmatrix} = \begin{bmatrix} \frac{1}{3} & -\frac{2}{3} \\ -\frac{2}{3} & \frac{7}{3} \end{bmatrix}$$

Check:

$$NN^{-1} = \begin{bmatrix} 7 & 2 \\ 2 & 1 \end{bmatrix} \cdot \begin{bmatrix} \frac{1}{3} & -\frac{2}{3} \\ -\frac{2}{3} & \frac{7}{3} \end{bmatrix} = \begin{bmatrix} \frac{7}{3} - \frac{4}{3} & -\frac{14}{3} + \frac{14}{3} \\ \frac{2}{3} - \frac{2}{3} & -\frac{4}{3} + \frac{7}{3} \end{bmatrix} = \begin{bmatrix} 1 & 0 \\ 0 & 1 \end{bmatrix}$$

$$N^{-1}N = \begin{bmatrix} \frac{1}{3} & -\frac{2}{3} \\ -\frac{2}{3} & \frac{7}{3} \end{bmatrix} \cdot \begin{bmatrix} 7 & 2 \\ 2 & 1 \end{bmatrix} = \begin{bmatrix} \frac{7}{3} - \frac{4}{3} & \frac{2}{3} - \frac{2}{3} \\ -\frac{14}{3} + \frac{14}{3} & -\frac{4}{3} + \frac{7}{3} \end{bmatrix} = \begin{bmatrix} 1 & 0 \\ 0 & 1 \end{bmatrix}$$

Exercises

Find the inverse of each matrix, if it exists.

1. $\begin{bmatrix} 24 & 12 \\ 8 & 4 \end{bmatrix}$

2. $\begin{bmatrix} 1 & 1 \\ 0 & 1 \end{bmatrix}$

3. $\begin{bmatrix} 40 & -10 \\ -20 & 30 \end{bmatrix}$

4. $\begin{bmatrix} 6 & 5 \\ 10 & 8 \end{bmatrix}$

5. $\begin{bmatrix} 3 & 6 \\ 4 & 8 \end{bmatrix}$

6. $\begin{bmatrix} 8 & 2 \\ 10 & 4 \end{bmatrix}$

4-8 Study Guide and Intervention

Using Matrices to Solve Systems of Equations

Write Matrix Equations A **matrix equation** for a system of equations consists of the product of the coefficient and variable matrices on the left and the constant matrix on the right of the equals sign.

Example Write a matrix equation for each system of equations.

a. $3x - 7y = 12$
$x + 5y = -8$

Determine the coefficient, variable, and constant matrices.

$$\begin{bmatrix} 3 & -7 \\ 1 & 5 \end{bmatrix} \cdot \begin{bmatrix} x \\ y \end{bmatrix} = \begin{bmatrix} 12 \\ -8 \end{bmatrix}$$

b. $2x - y + 3z = -7$
$x + 3y - 4z = 15$
$7x + 2y + z = -28$

$$\begin{bmatrix} 2 & -1 & 3 \\ 1 & 3 & -4 \\ 7 & 2 & 1 \end{bmatrix} \cdot \begin{bmatrix} x \\ y \\ z \end{bmatrix} = \begin{bmatrix} -7 \\ 15 \\ -28 \end{bmatrix}$$

Exercises

Write a matrix equation for each system of equations.

1. $2x + y = 8$
$5x - 3y = -12$

2. $4x - 3y = 18$
$x + 2y = 12$

3. $7x - 2y = 15$
$3x + y = -10$

4. $4x - 6y = 20$
$3x + y + 8 = 0$

5. $5x + 2y = 18$
$x = -4y + 25$

6. $3x - y = 24$
$3y = 80 - 2x$

7. $2x + y + 7z = 12$
$5x - y + 3z = 15$
$x + 2y - 6z = 25$

8. $5x - y + 7z = 32$
$x + 3y - 2z = -18$
$2x + 4y - 3z = 12$

9. $4x - 3y - z = -100$
$2x + y - 3z = -64$
$5x + 3y - 2z = 8$

10. $x - 3y + 7z = 27$
$2x + y - 5z = 48$
$4x - 2y + 3z = 72$

11. $2x + 3y - 9z = -108$
$x + 5z = 40 + 2y$
$3x + 5y = 89 + 4z$

12. $z = 45 - 3x + 2y$
$2x + 3y - z = 60$
$x = 4y - 2z + 120$

4-8 Study Guide and Intervention (continued)

Using Matrices to Solve Systems of Equations

Solve Systems of Equations Use inverse matrices to solve systems of equations written as matrix equations.

Solving Matrix Equations	If $AX = B$, then $X = A^{-1}B$, where A is the coefficient matrix, X is the variable matrix, and B is the constant matrix.

Example Solve $\begin{bmatrix} 5 & 2 \\ 6 & 4 \end{bmatrix} \begin{bmatrix} x \\ y \end{bmatrix} = \begin{bmatrix} 6 \\ 4 \end{bmatrix}$.

In the matrix equation $A = \begin{bmatrix} 5 & 2 \\ 6 & 4 \end{bmatrix}$, $X = \begin{bmatrix} x \\ y \end{bmatrix}$, and $B = \begin{bmatrix} 6 \\ 4 \end{bmatrix}$.

Step 1 Find the inverse of the coefficient matrix.

$$A^{-1} = \frac{1}{20 - 12} \begin{bmatrix} 4 & -2 \\ -6 & 5 \end{bmatrix} \text{ or } \frac{1}{8} \begin{bmatrix} 4 & -2 \\ -6 & 5 \end{bmatrix}.$$

Step 2 Multiply each side of the matrix equation by the inverse matrix.

$$\frac{1}{8} \begin{bmatrix} 4 & -2 \\ -6 & 5 \end{bmatrix} \cdot \begin{bmatrix} 5 & 2 \\ 6 & 4 \end{bmatrix} \cdot \begin{bmatrix} x \\ y \end{bmatrix} = \frac{1}{8} \begin{bmatrix} 4 & -2 \\ -6 & 5 \end{bmatrix} \cdot \begin{bmatrix} 6 \\ 4 \end{bmatrix} \qquad \text{Multiply each side by } A^{-1}.$$

$$\begin{bmatrix} 1 & 0 \\ 0 & 1 \end{bmatrix} \cdot \begin{bmatrix} x \\ y \end{bmatrix} = \frac{1}{8} \begin{bmatrix} 16 \\ -16 \end{bmatrix} \qquad \text{Multiply matrices.}$$

$$\begin{bmatrix} x \\ y \end{bmatrix} = \begin{bmatrix} 2 \\ -2 \end{bmatrix} \qquad \text{Simplify.}$$

The solution is $(2, -2)$.

Exercises

Solve each matrix equation or system of equations by using inverse matrices.

1. $\begin{bmatrix} 2 & 4 \\ 3 & -1 \end{bmatrix} \cdot \begin{bmatrix} x \\ y \end{bmatrix} = \begin{bmatrix} -2 \\ 18 \end{bmatrix}$

2. $\begin{bmatrix} -4 & -8 \\ 6 & 12 \end{bmatrix} \cdot \begin{bmatrix} x \\ y \end{bmatrix} = \begin{bmatrix} 16 \\ 12 \end{bmatrix}$

3. $\begin{bmatrix} 3 & 2 \\ 5 & 4 \end{bmatrix} \cdot \begin{bmatrix} x \\ y \end{bmatrix} = \begin{bmatrix} 3 \\ -7 \end{bmatrix}$

4. $\begin{bmatrix} 2 & -3 \\ 2 & 5 \end{bmatrix} \cdot \begin{bmatrix} x \\ y \end{bmatrix} = \begin{bmatrix} 4 \\ -8 \end{bmatrix}$

5. $\begin{bmatrix} 3 & 6 \\ 5 & 9 \end{bmatrix} \cdot \begin{bmatrix} x \\ y \end{bmatrix} = \begin{bmatrix} -15 \\ 6 \end{bmatrix}$

6. $\begin{bmatrix} 1 & 2 \\ 3 & -1 \end{bmatrix} \cdot \begin{bmatrix} x \\ y \end{bmatrix} = \begin{bmatrix} 3 \\ -6 \end{bmatrix}$

7. $4x - 2y = 22$
$6x + 4y = -2$

8. $2x - y = 2$
$x + 2y = 46$

9. $3x + 4y = 12$
$5x + 8y = -8$

10. $x + 3y = -5$
$2x + 7y = 8$

11. $5x + 4y = 5$
$9x - 8y = 0$

12. $3x - 2y = 5$
$x - 4y = 20$

5-1 Study Guide and Intervention

Graphing Quadratic Functions

Graph Quadratic Functions

Quadratic Function	A function defined by an equation of the form $f(x) = ax^2 + bx + c$, where $a \neq 0$
Graph of a Quadratic Function	A **parabola** with these characteristics: y intercept: c; axis of symmetry: $x = \frac{-b}{2a}$; x-coordinate of vertex: $\frac{-b}{2a}$

Example Find the y-intercept, the equation of the axis of symmetry, and the x-coordinate of the vertex for the graph of $f(x) = x^2 - 3x + 5$. Use this information to graph the function.

$a = 1$, $b = -3$, and $c = 5$, so the y-intercept is 5. The equation of the axis of symmetry is $x = \frac{-(-3)}{2(1)}$ or $\frac{3}{2}$. The x-coordinate of the vertex is $\frac{3}{2}$.

Next make a table of values for x near $\frac{3}{2}$.

x	$x^2 - 3x + 5$	$f(x)$	$(x, f(x))$
0	$0^2 - 3(0) + 5$	5	$(0, 5)$
1	$1^2 - 3(1) + 5$	3	$(1, 3)$
$\frac{3}{2}$	$\left(\frac{3}{2}\right)^2 - 3\left(\frac{3}{2}\right) + 5$	$\frac{11}{4}$	$\left(\frac{3}{2}, \frac{11}{4}\right)$
2	$2^2 - 3(2) + 5$	3	$(2, 3)$
3	$3^2 - 3(3) + 5$	5	$(3, 5)$

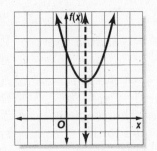

Exercises

For Exercises 1–3, complete parts a–c for each quadratic function.
a. Find the y-intercept, the equation of the axis of symmetry, and the x-coordinate of the vertex.
b. Make a table of values that includes the vertex.
c. Use this information to graph the function.

1. $f(x) = x^2 + 6x + 8$

2. $f(x) = -x^2 - 2x + 2$

3. $f(x) = 2x^2 - 4x + 3$

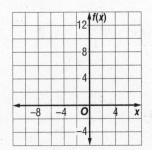

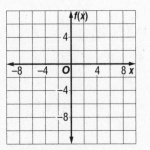

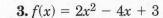

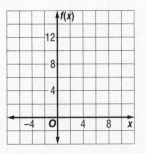

53

5-1 Study Guide and Intervention (continued)

Graphing Quadratic Functions

Maximum and Minimum Values The y-coordinate of the vertex of a quadratic function is the maximum or minimum value of the function.

Maximum or Minimum Value of a Quadratic Function	The graph of $f(x) = ax^2 + bx + c$, where $a \neq 0$, opens up and has a minimum when $a > 0$. The graph opens down and has a maximum when $a < 0$.

Example Determine whether each function has a maximum or minimum value, and find the maximum or minimum value of each function. Then state the domain and range of the function.

a. $f(x) = 3x^2 - 6x + 7$

For this function, $a = 3$ and $b = -6$.

Since $a > 0$, the graph opens up, and the function has a minimum value.

The minimum value is the y-coordinate of the vertex. The x-coordinate of the vertex is $\dfrac{-b}{2a} = -\dfrac{-6}{2(3)} = 1$.

Evaluate the function at $x = 1$ to find the minimum value.

$f(1) = 3(1)^2 - 6(1) + 7 = 4$, so the minimum value of the function is 4. The domain is all real numbers. The range is all reals greater than or equal to the minimum value, that is $\{f(x) \mid f(x) \geq 4\}$.

b. $f(x) = 100 - 2x - x^2$

For this function, $a = -1$ and $b = -2$.

Since $a < 0$, the graph opens down, and the function has a maximum value.

The maximum value is the y-coordinate of the vertex. The x-coordinate of the vertex is $\dfrac{-b}{2a} = -\dfrac{-2}{2(-1)} = -1$.

Evaluate the function at $x = -1$ to find the maximum value.

$f(-1) = 100 - 2(-1) - (-1)^2 = 101$, so the minimum value of the function is 101. The domain is all real numbers. The range is all reals less than or equal to the maximum value, that is $\{f(x) \mid f(x) \leq 101\}$.

Exercises

Determine whether each function has a maximum or minimum value, and find the maximum or minimum value. Then state the domain and range of the function.

1. $f(x) = 2x^2 - x + 10$

2. $f(x) = x^2 + 4x - 7$

3. $f(x) = 3x^2 - 3x + 1$

4. $f(x) = 16 + 4x - x^2$

5. $f(x) = x^2 - 7x + 11$

6. $f(x) = -x^2 + 6x - 4$

7. $f(x) = x^2 + 5x + 2$

8. $f(x) = 20 + 6x - x^2$

9. $f(x) = 4x^2 + x + 3$

10. $f(x) = -x^2 - 4x + 10$

11. $f(x) = x^2 - 10x + 5$

12. $f(x) = -6x^2 + 12x + 21$

NAME _____ DATE _____ PERIOD _____

5-2 Study Guide and Intervention
Solving Quadratic Equations by Graphing

Solve Quadratic Equations

Quadratic Equation	A quadratic equation has the form $ax^2 + bx + c = 0$, where $a \neq 0$.
Roots of a Quadratic Equation	solution(s) of the equation, or the zero(s) of the related quadratic function

The zeros of a quadratic function are the x-intercepts of its graph. Therefore, finding the x-intercepts is one way of solving the related quadratic equation.

Example Solve $x^2 + x - 6 = 0$ by graphing.

Graph the related function $f(x) = x^2 + x - 6$.

The x-coordinate of the vertex is $\frac{-b}{2a} = -\frac{1}{2}$, and the equation of the axis of symmetry is $x = -\frac{1}{2}$.

Make a table of values using x-values around $-\frac{1}{2}$.

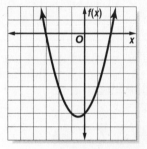

x	-1	$-\frac{1}{2}$	0	1	2
$f(x)$	-6	$-6\frac{1}{4}$	-6	-4	0

From the table and the graph, we can see that the zeros of the function are 2 and -3.

Exercises

Solve each equation by graphing.

1. $x^2 + 2x - 8 = 0$

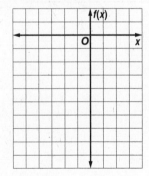

2. $x^2 - 4x - 5 = 0$

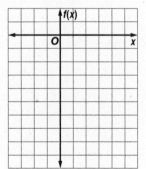

3. $x^2 - 5x + 4 = 0$

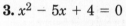
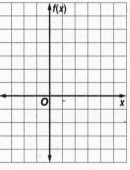

4. $x^2 - 10x + 21 = 0$

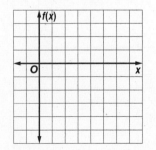

5. $x^2 + 4x + 6 = 0$

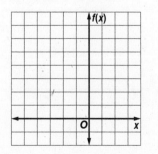

6. $4x^2 + 4x + 1 = 0$

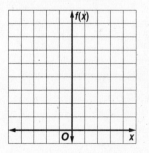

5-2 Study Guide and Intervention (continued)

Solving Quadratic Equations by Graphing

Estimate Solutions Often, you may not be able to find exact solutions to quadratic equations by graphing. But you can use the graph to estimate solutions.

Example Solve $x^2 - 2x - 2 = 0$ by graphing. If exact roots cannot be found, state the consecutive integers between which the roots are located.

The equation of the axis of symmetry of the related function is $x = -\dfrac{-2}{2(1)} = 1$, so the vertex has x-coordinate 1. Make a table of values.

x	−1	0	1	2	3
$f(x)$	1	−2	−3	−2	1

The x-intercepts of the graph are between 2 and 3 and between 0 and −1. So one solution is between 2 and 3, and the other solution is between 0 and −1.

Exercises

Solve the equations by graphing. If exact roots cannot be found, state the consecutive integers between which the roots are located.

1. $x^2 - 4x + 2 = 0$

2. $x^2 + 6x + 6 = 0$

3. $x^2 + 4x + 2 = 0$

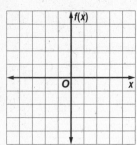

4. $-x^2 + 2x + 4 = 0$

5. $2x^2 - 12x + 17 = 0$

6. $-\dfrac{1}{2}x^2 + x + \dfrac{5}{2} = 0$

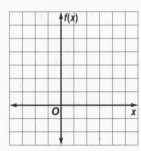
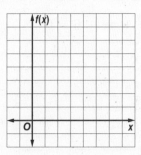
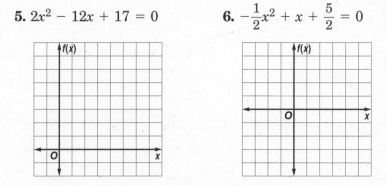
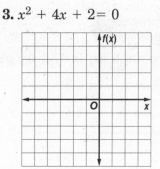

5-3 Study Guide and Intervention

Solving Quadratic Equations by Factoring

Solve Equations by Factoring When you use factoring to solve a quadratic equation, you use the following property.

Zero Product Property	For any real numbers a and b, if $ab = 0$, then either $a = 0$ or $b = 0$, or both a and $b = 0$.

Example Solve each equation by factoring.

a. $3x^2 = 15x$

$3x^2 = 15x$ Original equation

$3x^2 - 15x = 0$ Subtract 15x from both sides.

$3x(x - 5) = 0$ Factor the binomial.

$3x = 0$ or $x - 5 = 0$ Zero Product Property

$x = 0$ or $x = 5$ Solve each equation.

The solution set is {0, 5}.

b. $4x^2 - 5x = 21$

$4x^2 - 5x = 21$ Original equation

$4x^2 - 5x - 21 = 0$ Subtract 21 from both sides.

$(4x + 7)(x - 3) = 0$ Factor the trinomial.

$4x + 7 = 0$ or $x - 3 = 0$ Zero Product Property

$x = -\dfrac{7}{4}$ or $x = 3$ Solve each equation.

The solution set is $\left\{-\dfrac{7}{4}, 3\right\}$.

Exercises

Solve each equation by factoring.

1. $6x^2 - 2x = 0$

2. $x^2 = 7x$

3. $20x^2 = -25x$

4. $6x^2 = 7x$

5. $6x^2 - 27x = 0$

6. $12x^2 - 8x = 0$

7. $x^2 + x - 30 = 0$

8. $2x^2 - x - 3 = 0$

9. $x^2 + 14x + 33 = 0$

10. $4x^2 + 27x - 7 = 0$

11. $3x^2 + 29x - 10 = 0$

12. $6x^2 - 5x - 4 = 0$

13. $12x^2 - 8x + 1 = 0$

14. $5x^2 + 28x - 12 = 0$

15. $2x^2 - 250x + 5000 = 0$

16. $2x^2 - 11x - 40 = 0$

17. $2x^2 + 21x - 11 = 0$

18. $3x^2 + 2x - 21 = 0$

19. $8x^2 - 14x + 3 = 0$

20. $6x^2 + 11x - 2 = 0$

21. $5x^2 + 17x - 12 = 0$

22. $12x^2 + 25x + 12 = 0$

23. $12x^2 + 18x + 6 = 0$

24. $7x^2 - 36x + 5 = 0$

5-3 Study Guide and Intervention (continued)

Solving Quadratic Equations by Factoring

Write Quadratic Equations To write a quadratic equation with roots p and q, let $(x - p)(x - q) = 0$. Then multiply using FOIL.

Example Write a quadratic equation with the given roots. Write the equation in standard form.

a. 3, −5

$(x - p)(x - q) = 0$	Write the pattern.
$(x - 3)[x - (-5)] = 0$	Replace p with 3, q with −5.
$(x - 3)(x + 5) = 0$	Simplify.
$x^2 + 2x - 15 = 0$	Use FOIL.

The equation $x^2 + 2x - 15 = 0$ has roots 3 and −5.

b. $-\dfrac{7}{8}, \dfrac{1}{3}$

$$(x - p)(x - q) = 0$$

$$\left[x - \left(-\frac{7}{8}\right)\right]\left(x - \frac{1}{3}\right) = 0$$

$$\left(x + \frac{7}{8}\right)\left(x - \frac{1}{3}\right) = 0$$

$$\frac{(8x + 7)}{8} \cdot \frac{(3x - 1)}{3} = 0$$

$$\frac{24 \cdot (8x + 7)(3x - 1)}{24} = 24 \cdot 0$$

$$24x^2 + 13x - 7 = 0$$

The equation $24x^2 + 13x - 7 = 0$ has roots $-\dfrac{7}{8}$ and $\dfrac{1}{3}$.

Exercises

Write a quadratic equation with the given roots. Write the equation in standard form.

1. $3, -4$

2. $-8, -2$

3. $1, 9$

4. -5

5. $10, 7$

6. $-2, 15$

7. $-\dfrac{1}{3}, 5$

8. $2, \dfrac{2}{3}$

9. $-7, \dfrac{3}{4}$

10. $3, \dfrac{2}{5}$

11. $-\dfrac{4}{9}, -1$

12. $9, \dfrac{1}{6}$

13. $\dfrac{2}{3}, -\dfrac{2}{3}$

14. $\dfrac{5}{4}, -\dfrac{1}{2}$

15. $\dfrac{3}{7}, \dfrac{1}{5}$

16. $-\dfrac{7}{8}, \dfrac{7}{2}$

17. $\dfrac{1}{2}, \dfrac{3}{4}$

18. $\dfrac{1}{8}, \dfrac{1}{6}$

NAME _____ DATE _____ PERIOD _____

5-4 Study Guide and Intervention
Complex Numbers

SQUARE ROOTS A **square root** of a number n is a number whose square is n. For nonnegative real numbers a and b, $\sqrt{ab} = \sqrt{a} \cdot \sqrt{b}$ and $\sqrt{\frac{a}{b}} = \frac{\sqrt{a}}{\sqrt{b}}, b \neq 0$. The **imaginary unit i** is defined to have the property that $i^2 = -1$. Simplified square root expressions do not have radicals in the denominator, and any number remaining under the square root has no perfect square factor other than 1.

Example 1

a. **Simplify** $\sqrt{48}$.
$$\sqrt{48} = \sqrt{16 \cdot 3}$$
$$= \sqrt{16} \cdot \sqrt{3}$$
$$= 4\sqrt{3}$$

b. **Simplify** $\sqrt{-63}$.
$$\sqrt{-63} = \sqrt{-1 \cdot 7 \cdot 9}$$
$$= \sqrt{-1} \cdot \sqrt{7} \cdot \sqrt{9}$$
$$= 3i\sqrt{7}$$

Example 2

a. **Simplify** $\sqrt{125x^2y^5}$.
$$\sqrt{125x^2y^5} = \sqrt{5 \cdot 25x^2y^4y}$$
$$= \sqrt{25} \cdot \sqrt{5} \cdot \sqrt{x^2} \cdot \sqrt{y^4} \cdot \sqrt{y}$$
$$= 5xy^2\sqrt{5y}$$

b. **Simplify** $\sqrt{-44x^6}$.
$$\sqrt{-44x^6} = \sqrt{-1 \cdot 4 \cdot 11 \cdot x^6}$$
$$= \sqrt{-1} \cdot \sqrt{4} \cdot \sqrt{11} \cdot \sqrt{x^6}$$
$$= 2i\sqrt{11}x^3$$

Example 3 Solve $x^2 + 5 = 0$.

$x^2 + 5 = 0$	Original equation.
$x^2 = -5$	Subtract 5 from each side.
$x = \pm\sqrt{5}i$	Square Root Property.

Exercises

Simplify.

1. $\sqrt{72}$

2. $\sqrt{-24}$

3. $\sqrt{\frac{128}{147}}$

4. $\sqrt{75x^4y^7}$

5. $\sqrt{-84}$

6. $\sqrt{-32xy^4}$

Solve each equation.

7. $5x^2 + 45 = 0$

8. $4x^2 + 24 = 0$

9. $-9x^2 = 9$

10. $7x^2 + 84 = 0$

Study Guide and Intervention   59   Glencoe Algebra 2

5-4 Study Guide and Intervention (continued)

Complex Numbers

Operations with Complex Numbers

Complex Number	A complex number is any number that can be written in the form $a + bi$, where a and b are real numbers and i is the imaginary unit ($i^2 = -1$). a is called the real part, and b is called the imaginary part.
Addition and Subtraction of Complex Numbers	Combine like terms. $(a + bi) + (c + di) = (a + c) + (b + d)i$ $(a + bi) - (c + di) = (a - c) + (b - d)i$
Multiplication of Complex Numbers	Use the definition of i^2 and the FOIL method: $(a + bi)(c + di) = (ac - bd) + (ad + bc)i$
Complex Conjugate	$a + bi$ and $a - bi$ are complex conjugates. The product of complex conjugates is always a real number.

To divide by a complex number, first multiply the dividend and divisor by the **complex conjugate** of the divisor.

Example 1 Simplify $(6 + i) + (4 - 5i)$.

$(6 + i) + (4 - 5i)$
$= (6 + 4) + (1 - 5)i$
$= 10 - 4i$

Example 2 Simplify $(8 + 3i) - (6 - 2i)$.

$(8 + 3i) - (6 - 2i)$
$= (8 - 6) + [3 - (-2)]i$
$= 2 + 5i$

Example 3 Simplify $(2 - 5i) \cdot (-4 + 2i)$.

$(2 - 5i) \cdot (-4 + 2i)$
$= 2(-4) + 2(2i) + (-5i)(-4) + (-5i)(2i)$
$= -8 + 4i + 20i - 10i^2$
$= -8 + 24i - 10(-1)$
$= 2 + 24i$

Example 4 Simplify $\dfrac{3 - i}{2 + 3i}$.

$\dfrac{3 - i}{2 + 3i} = \dfrac{3 - i}{2 + 3i} \cdot \dfrac{2 - 3i}{2 - 3i}$

$= \dfrac{6 - 9i - 2i + 3i^2}{4 - 9i^2}$

$= \dfrac{3 - 11i}{13}$

$= \dfrac{3}{13} - \dfrac{11}{13}i$

Exercises

Simplify.

1. $(-4 + 2i) + (6 - 3i)$

2. $(5 - i) - (3 - 2i)$

3. $(6 - 3i) + (4 - 2i)$

4. $(-11 + 4i) - (1 - 5i)$

5. $(8 + 4i) + (8 - 4i)$

6. $(5 + 2i) - (-6 - 3i)$

7. $(2 + i)(3 - i)$

8. $(5 - 2i)(4 - i)$

9. $(4 - 2i)(1 - 2i)$

10. $\dfrac{5}{3 + i}$

11. $\dfrac{7 - 13i}{2i}$

12. $\dfrac{6 - 5i}{3i}$

5-5 Study Guide and Intervention

Completing the Square

Square Root Property Use the Square Root Property to solve a quadratic equation that is in the form "perfect square trinomial = constant."

Example Solve each equation by using the Square Root Property.

a. $x^2 - 8x + 16 = 25$

$$x^2 - 8x + 16 = 25$$
$$(x - 4)^2 = 25$$
$$x - 4 = \sqrt{25} \quad \text{or} \quad x - 4 = -\sqrt{25}$$
$$x = 5 + 4 = 9 \text{ or} \quad x = -5 + 4 = -1$$

The solution set is $\{9, -1\}$.

b. $4x^2 - 20x + 25 = 32$

$$4x^2 - 20x + 25 = 32$$
$$(2x - 5)^2 = 32$$
$$2x - 5 = \sqrt{32} \text{ or } 2x - 5 = -\sqrt{32}$$
$$2x - 5 = 4\sqrt{2} \text{ or } 2x - 5 = -4\sqrt{2}$$
$$x = \frac{5 \pm 4\sqrt{2}}{2}$$

The solution set is $\left\{\dfrac{5 \pm 4\sqrt{2}}{2}\right\}$.

Exercises

Solve each equation by using the Square Root Property.

1. $x^2 - 18x + 81 = 49$

2. $x^2 + 20x + 100 = 64$

3. $4x^2 + 4x + 1 = 16$

4. $36x^2 + 12x + 1 = 18$

5. $9x^2 - 12x + 4 = 4$

6. $25x^2 + 40x + 16 = 28$

7. $4x^2 - 28x + 49 = 64$

8. $16x^2 + 24x + 9 = 81$

9. $100x^2 - 60x + 9 = 121$

10. $25x^2 + 20x + 4 = 75$

11. $36x^2 + 48x + 16 = 12$

12. $25x^2 - 30x + 9 = 96$

5-5 Study Guide and Intervention (continued)

Completing the Square

Complete the Square To complete the square for a quadratic expression of the form $x^2 + bx$, follow these steps.

1. Find $\dfrac{b}{2}$. → 2. Square $\dfrac{b}{2}$. → 3. Add $\left(\dfrac{b}{2}\right)^2$ to $x^2 + bx$.

Example 1 Find the value of c that makes $x^2 + 22x + c$ a perfect square trinomial. Then write the trinomial as the square of a binomial.

Step 1 $b = 22$; $\dfrac{b}{2} = 11$

Step 2 $11^2 = 121$

Step 3 $c = 121$

The trinomial is $x^2 + 22x + 121$, which can be written as $(x + 11)^2$.

Example 2 Solve $2x^2 - 8x - 24 = 0$ by completing the square.

$2x^2 - 8x - 24 = 0$	Original equation
$\dfrac{2x^2 - 8x - 24}{2} = \dfrac{0}{2}$	Divide each side by 2.
$x^2 - 4x - 12 = 0$	$x^2 - 4x - 12$ is not a perfect square.
$x^2 - 4x = 12$	Add 12 to each side.
$x^2 - 4x + 4 = 12 + 4$	Since $\left(-\dfrac{4}{2}\right)^2 = 4$, add 4 to each side.
$(x - 2)^2 = 16$	Factor the square.
$x - 2 = \pm 4$	Square Root Property
$x = 6$ or $x = -2$	Solve each equation.

The solution set is {6, −2}.

Exercises

Find the value of c that makes each trinomial a perfect square. Then write the trinomial as a perfect square.

1. $x^2 - 10x + c$

2. $x^2 + 60x + c$

3. $x^2 - 3x + c$

4. $x^2 + 3.2x + c$

5. $x^2 + \dfrac{1}{2}x + c$

6. $x^2 - 2.5x + c$

Solve each equation by completing the square.

7. $y^2 - 4y - 5 = 0$

8. $x^2 - 8x - 65 = 0$

9. $s^2 - 10s + 21 = 0$

10. $2x^2 - 3x + 1 = 0$

11. $2x^2 - 13x - 7 = 0$

12. $25x^2 + 40x - 9 = 0$

13. $x^2 + 4x + 1 = 0$

14. $y^2 + 12y + 4 = 0$

15. $t^2 + 3t - 8 = 0$

5-6 Study Guide and Intervention

The Quadratic Formula and the Discriminant

Quadratic Formula The **Quadratic Formula** can be used to solve *any* quadratic equation once it is written in the form $ax^2 + bx + c = 0$.

Quadratic Formula	The solutions of $ax^2 + bx + c = 0$, with $a \neq 0$, are given by $x = \dfrac{-b \pm \sqrt{b^2 - 4ac}}{2a}$.

Example   Solve $x^2 - 5x = 14$ by using the Quadratic Formula.

Rewrite the equation as $x^2 - 5x - 14 = 0$.

$x = \dfrac{-b \pm \sqrt{b^2 - 4ac}}{2a}$ Quadratic Formula

$= \dfrac{-(-5) \pm \sqrt{(-5)^2 - 4(1)(-14)}}{2(1)}$ Replace a with 1, b with -5, and c with -14.

$= \dfrac{5 \pm \sqrt{81}}{2}$ Simplify.

$= \dfrac{5 \pm 9}{2}$

$= 7 \text{ or } -2$

The solutions are -2 and 7.

Exercises

Solve each equation by using the Quadratic Formula.

1. $x^2 + 2x - 35 = 0$ **2.** $x^2 + 10x + 24 = 0$ **3.** $x^2 - 11x + 24 = 0$

4. $4x^2 + 19x - 5 = 0$ **5.** $14x^2 + 9x + 1 = 0$ **6.** $2x^2 - x - 15 = 0$

7. $3x^2 + 5x = 2$ **8.** $2y^2 + y - 15 = 0$ **9.** $3x^2 - 16x + 16 = 0$

10. $8x^2 + 6x - 9 = 0$ **11.** $r^2 - \dfrac{3r}{5} + \dfrac{2}{25} = 0$ **12.** $x^2 - 10x - 50 = 0$

13. $x^2 + 6x - 23 = 0$ **14.** $4x^2 - 12x - 63 = 0$ **15.** $x^2 - 6x + 21 = 0$

5-6 Study Guide and Intervention (continued)
The Quadratic Formula and the Discriminant

Roots and the Discriminant

Discriminant	The expression under the radical sign, $b^2 - 4ac$, in the Quadratic Formula is called the **discriminant**.

Roots of a Quadratic Equation

Discriminant	Type and Number of Roots
$b^2 - 4ac > 0$ and a perfect square	2 rational roots
$b^2 - 4ac > 0$, but **not** a perfect square	2 irrational roots
$b^2 - 4ac = 0$	1 rational root
$b^2 - 4ac < 0$	2 complex roots

Example Find the value of the discriminant for each equation. Then describe the number and types of roots for the equation.

a. $2x^2 + 5x + 3$

The discriminant is
$b^2 - 4ac = 5^2 - 4(2)(3)$ or 1.
The discriminant is a perfect square, so the equation has 2 rational roots.

b. $3x^2 - 2x + 5$

The discriminant is
$b^2 - 4ac = (-2)^2 - 4(3)(5)$ or -56.
The discriminant is negative, so the equation has 2 complex roots.

Exercises

For Exercises 1–12, complete parts a–c for each quadratic equation.
a. Find the value of the discriminant.
b. Describe the number and type of roots.
c. Find the exact solutions by using the Quadratic Formula.

1. $p^2 + 12p = -4$

2. $9x^2 - 6x + 1 = 0$

3. $2x^2 - 7x - 4 = 0$

4. $x^2 + 4x - 4 = 0$

5. $5x^2 - 36x + 7 = 0$

6. $4x^2 - 4x + 11 = 0$

7. $x^2 - 7x + 6 = 0$

8. $m^2 - 8m = -14$

9. $25x^2 - 40x = -16$

10. $4x^2 + 20x + 29 = 0$

11. $6x^2 + 26x + 8 = 0$

12. $4x^2 - 4x - 11 = 0$

5-7 Study Guide and Intervention

Analyzing Graphs of Quadratic Functions

Analyze Quadratic Functions

	The graph of $y = a(x - h)^2 + k$ has the following characteristics:
Vertex Form of a Quadratic Function	• Vertex: (h, k) • Axis of symmetry: $x = h$ • Opens up if $a > 0$ • Opens down if $a < 0$ • Narrower than the graph of $y = x^2$ if $|a| > 1$ • Wider than the graph of $y = x^2$ if $|a| < 1$

Example Identify the vertex, axis of symmetry, and direction of opening of each graph.

a. $y = 2(x + 4)^2 - 11$

The vertex is at (h, k) or $(-4, -11)$, and the axis of symmetry is $x = -4$. The graph opens up.

b. $y = -\frac{1}{4}(x - 2)^2 + 10$

The vertex is at (h, k) or $(2, 10)$, and the axis of symmetry is $x = 2$. The graph opens down.

Exercises

Each quadratic function is given in vertex form. Identify the vertex, axis of symmetry, and direction of opening of the graph.

1. $y = (x - 2)^2 + 16$ **2.** $y = 4(x + 3)^2 - 7$ **3.** $y = \frac{1}{2}(x - 5)^2 + 3$

4. $y = -7(x + 1)^2 - 9$ **5.** $y = \frac{1}{5}(x - 4)^2 - 12$ **6.** $y = 6(x + 6)^2 + 6$

7. $y = \frac{2}{5}(x - 9)^2 + 12$ **8.** $y = 8(x - 3)^2 - 2$ **9.** $y = -3(x - 1)^2 - 2$

10. $y = -\frac{5}{2}(x + 5)^2 + 12$ **11.** $y = \frac{4}{3}(x - 7)^2 + 22$ **12.** $y = 16(x - 4)^2 + 1$

13. $y = 3(x - 1.2)^2 + 2.7$ **14.** $y = -0.4(x - 0.6)^2 - 0.2$ **15.** $y = 1.2(x + 0.8)^2 + 6.5$

5-7 Study Guide and Intervention (continued)

Analyzing Graphs of Quadratic Functions

Write Quadratic Functions in Vertex Form A quadratic function is easier to graph when it is in vertex form. You can write a quadratic function of the form $y = ax^2 + bx + c$ in vertex from by completing the square.

Example Write $y = 2x^2 - 12x + 25$ in vertex form. Then graph the function.

$y = 2x^2 - 12x + 25$

$y = 2(x^2 - 6x) + 25$

$y = 2(x^2 - 6x + 9) + 25 - 18$

$y = 2(x - 3)^2 + 7$

The vertex form of the equation is $y = 2(x - 3)^2 + 7$.

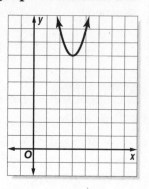

Exercises

Write each quadratic function in vertex form. Then graph the function.

1. $y = x^2 - 10x + 32$ **2.** $y = x^2 + 6x$ **3.** $y = x^2 - 8x + 6$

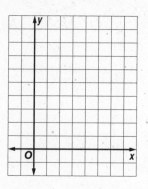

4. $y = -4x^2 + 16x - 11$ **5.** $y = 3x^2 - 12x + 5$ **6.** $y = 5x^2 - 10x + 9$

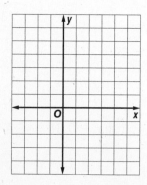
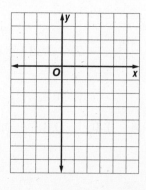
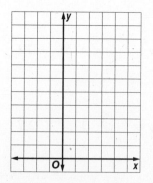

5-8 Study Guide and Intervention

Graphing and Solving Quadratic Inequalities

Graph Quadratic Inequalities To graph a quadratic inequality in two variables, use the following steps:

1. Graph the related quadratic equation, $y = ax^2 + bx + c$.
   Use a dashed line for $<$ or $>$; use a solid line for $\leq$ or $\geq$.

2. Test a point inside the parabola.
   If it satisfies the inequality, shade the region inside the parabola; otherwise, shade the region outside the parabola.

Example Graph the inequality $y > x^2 + 6x + 7$.

First graph the equation $y = x^2 + 6x + 7$. By completing the square, you get the vertex form of the equation $y = (x + 3)^2 - 2$, so the vertex is $(-3, -2)$. Make a table of values around $x = -3$, and graph. Since the inequality includes $>$, use a dashed line.

Test the point $(-3, 0)$, which is inside the parabola. Since $(-3)^2 + 6(-3) + 7 = -2$, and $0 > -2$, $(-3, 0)$ satisfies the inequality. Therefore, shade the region inside the parabola.

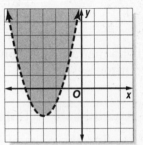

Exercises

Graph each inequality.

1. $y > x^2 - 8x + 17$

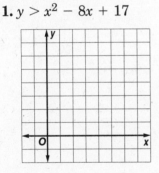

2. $y \leq x^2 + 6x + 4$

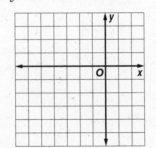

3. $y \geq x^2 + 2x + 2$

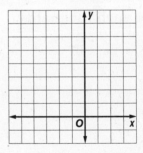

4. $y < -x^2 + 4x - 6$

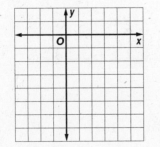

5. $y \geq 2x^2 + 4x$

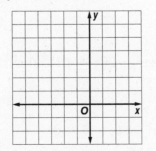

6. $y > -2x^2 - 4x + 2$

5-8 Study Guide and Intervention (continued)

Graphing and Solving Quadratic Inequalities

Solve Quadratic Inequalities Quadratic inequalities in one variable can be solved graphically or algebraically.

Graphical Method	To solve $ax^2 + bx + c < 0$: First graph $y = ax^2 + bx + c$. The solution consists of the x-values for which the graph is **below** the x-axis. To solve $ax^2 + bx + c > 0$: First graph $y = ax^2 + bx + c$. The solution consists the x-values for which the graph is **above** the x-axis.
Algebraic Method	Find the roots of the related quadratic equation by factoring, completing the square, or using the Quadratic Formula. 2 roots divide the number line into 3 intervals. Test a value in each interval to see which intervals are solutions.

If the inequality involves $\leq$ or $\geq$, the roots of the related equation are included in the solution set.

Example   Solve the inequality $x^2 - x - 6 \leq 0$.

First find the roots of the related equation $x^2 - x - 6 = 0$. The equation factors as $(x - 3)(x + 2) = 0$, so the roots are 3 and -2. The graph opens up with x-intercepts 3 and -2, so it must be on or below the x-axis for $-2 \leq x \leq 3$. Therefore the solution set is $\{x \mid -2 \leq x \leq 3\}$.

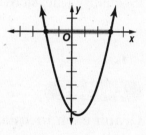

Exercises

Solve each inequality.

1. $x^2 + 2x < 0$

2. $x^2 - 16 < 0$

3. $0 < 6x - x^2 - 5$

4. $c^2 \leq 4$

5. $2m^2 - m < 1$

6. $y^2 < -8$

7. $x^2 - 4x - 12 < 0$

8. $x^2 + 9x + 14 > 0$

9. $-x^2 + 7x - 10 \geq 0$

10. $2x^2 + 5x - 3 \leq 0$

11. $4x^2 - 23x + 15 > 0$

12. $-6x^2 - 11x + 2 < 0$

13. $2x^2 - 11x + 12 \geq 0$

14. $x^2 - 4x + 5 < 0$

15. $3x^2 - 16x + 5 < 0$

6-1 Study Guide and Intervention

Properties of Exponents

Multiply and Divide Monomials Negative exponents are a way of expressing the multiplicative inverse of a number.

Negative Exponents	$a^{-n} = \dfrac{1}{a^n}$ and $\dfrac{1}{a^{-n}} = a^n$ for any real number $a \neq 0$ and any integer n.

When you **simplify an expression**, you rewrite it without parentheses or negative exponents. The following properties are useful when simplifying expressions.

Product of Powers	$a^m \cdot a^n = a^{m+n}$ for any real number a and integers m and n.
Quotient of Powers	$\dfrac{a^m}{a^n} = a^{m-n}$ for any real number $a \neq 0$ and integers m and n.
Properties of Powers	For a, b real numbers and m, n integers: $(a^m)^n = a^{mn}$ $(ab)^m = a^m b^m$ $\left(\dfrac{a}{b}\right)^n = \dfrac{a^n}{b^n}, b \neq 0$ $\left(\dfrac{a}{b}\right)^{-n} = \left(\dfrac{b}{a}\right)^n$ or $\dfrac{b^n}{a^n}, a \neq 0, b \neq 0$

Example Simplify. Assume that no variable equals 0.

a. $(3m^4n^{-2})(-5mn)^2$

$(3m^4n^{-2})(-5mn)^2 = 3m^4n^{-2} \cdot 25m^2n^2$
$= 75m^4m^2n^{-2}n^2$
$= 75m^{4+2}n^{-2+2}$
$= 75m^6$

b. $\dfrac{(-m^4)^3}{(2m^2)^{-2}}$

$\dfrac{(-m^4)^3}{(2m^2)^{-2}} = \dfrac{-m^{12}}{\dfrac{1}{4m^4}}$
$= -m^{12} \cdot 4m^4$
$= -4m^{16}$

Exercises

Simplify. Assume that no variable equals 0.

1. $c^{12} \cdot c^{-4} \cdot c^6$

2. $\dfrac{b^8}{b^2}$

3. $(a^4)^5$

4. $\dfrac{x^{-2}y}{x^4y^{-1}}$

5. $\left(\dfrac{a^2b}{a^{-3}b^2}\right)^{-1}$

6. $\left(\dfrac{x^2y}{xy^3}\right)^2$

7. $\dfrac{1}{5}(-5a^2b^3)^2(abc)^2$

8. $m^7 \cdot m^8$

9. $\dfrac{8m^3n^2}{4mn^3}$

10. $\dfrac{2^3c^4t^2}{2^2c^4t^2}$

11. $4j(2j^{-2}k^2)(3j^3k^{-7})$

12. $\dfrac{2mn^2(3m^2n)^2}{12m^3n^4}$

6-1 Study Guide and Intervention (continued)

Properties of Exponents

Scientific Notation

Scientific notation	A number expressed in the form $a \times 10^n$, where $1 \le a < 10$ and n is an integer

Example 1 **Express 46,000,000 in scientific notation.**

$46,000,000 = 4.6 \times 10,000,000$ $1 \le 4.6 < 10$

$\qquad\qquad = 4.6 \times 10^7$ Write 10,000,000 as a power of ten.

Example 2 **Evaluate $\dfrac{3.5 \times 10^4}{5 \times 10^{-2}}$. Express the result in scientific notation.**

$\dfrac{3.5 \times 10^4}{5 \times 10^{-2}} = \dfrac{3.5}{5} \times \dfrac{10^4}{10^{-2}}$

$\qquad\qquad = 0.7 \times 10^6$

$\qquad\qquad = 7 \times 10^5$

Exercises

Express each number in scientific notation.

1. 24,300

2. 0.00099

3. 4,860,000

4. 525,000,000

5. 0.0000038

6. 221,000

7. 0.000000064

8. 16,750

9. 0.000369

Evaluate. Express the result in scientific notation.

10. $(3.6 \times 10^4)(5 \times 10^3)$

11. $(1.4 \times 10^{-8})(8 \times 10^{12})$

12. $(4.2 \times 10^{-3})(3 \times 10^{-2})$

13. $\dfrac{9.5 \times 10^7}{3.8 \times 10^{-2}}$

14. $\dfrac{1.62 \times 10^{-2}}{1.8 \times 10^5}$

15. $\dfrac{4.81 \times 10^8}{6.5 \times 10^4}$

16. $(3.2 \times 10^{-3})^2$

17. $(4.5 \times 10^7)^2$

18. $(6.8 \times 10^{-5})^2$

19. ASTRONOMY Pluto is 3,674.5 million miles from the sun. Write this number in scientific notation.

20. CHEMISTRY The boiling point of the metal tungsten is 10,220°F. Write this temperature in scientific notation.

21. BIOLOGY The human body contains 0.0004% iodine by weight. How many pounds of iodine are there in a 120-pound teenager? Express your answer in scientific notation.

6-2 Study Guide and Intervention

Operations with Polynomials

Add and Subtract Polynomials

Polynomial	a monomial or a sum of monomials
Like Terms	terms that have the same variable(s) raised to the same power(s)

To add or subtract polynomials, perform the indicated operations and combine like terms.

Example 1 Simplify $-6rs + 18r^2 - 5s^2 - 14r^2 + 8rs - 6s^2$.

$-6rs + 18r^2 - 5s^2 - 14r^2 + 8rs - 6s^2$
$= (18r^2 - 14r^2) + (-6rs + 8rs) + (-5s^2 - 6s^2)$ Group like terms.
$= 4r^2 + 2rs - 11s^2$ Combine like terms.

Example 2 Simplify $4xy^2 + 12xy - 7x^2y - (20xy + 5xy^2 - 8x^2y)$.

$4xy^2 + 12xy - 7x^2y - (20xy + 5xy^2 - 8x^2y)$
$= 4xy^2 + 12xy - 7x^2y - 20xy - 5xy^2 + 8x^2y$ Distribute the minus sign.
$= (-7x^2y + 8x^2y) + (4xy^2 - 5xy^2) + (12xy - 20xy)$ Group like terms.
$= x^2y - xy^2 - 8xy$ Combine like terms.

Exercises

Simplify.

1. $(6x^2 - 3x + 2) - (4x^2 + x - 3)$

2. $(7y^2 + 12xy - 5x^2) + (6xy - 4y^2 - 3x^2)$

3. $(-4m^2 - 6m) - (6m + 4m^2)$

4. $27x^2 - 5y^2 + 12y^2 - 14x^2$

5. $(18p^2 + 11pq - 6q^2) - (15p^2 - 3pq + 4q^2)$

6. $17j^2 - 12k^2 + 3j^2 - 15j^2 + 14k^2$

7. $(8m^2 - 7n^2) - (n^2 - 12m^2)$

8. $14bc + 6b - 4c + 8b - 8c + 8bc$

9. $6r^2s + 11rs^2 + 3r^2s - 7rs^2 + 15r^2s - 9rs^2$

10. $-9xy + 11x^2 - 14y^2 - (6y^2 - 5xy - 3x^2)$

11. $(12xy - 8x + 3y) + (15x - 7y - 8xy)$

12. $10.8b^2 - 5.7b + 7.2 - (2.9b^2 - 4.6b - 3.1)$

13. $(3bc - 9b^2 - 6c^2) + (4c^2 - b^2 + 5bc)$

14. $11x^2 + 4y^2 + 6xy + 3y^2 - 5xy - 10x^2$

15. $\frac{1}{4}x^2 - \frac{3}{8}xy + \frac{1}{2}y^2 - \frac{1}{2}xy + \frac{1}{4}y^2 - \frac{3}{8}x^2$

16. $24p^3 - 15p^2 + 3p - 15p^3 + 13p^2 - 7p$

6-2 Study Guide and Intervention (continued)

Operations with Polynomials

Multiply Polynomials You use the distributive property when you multiply polynomials. When multiplying binomials, the **FOIL** pattern is helpful.

FOIL Pattern	To multiply two binomials, add the products of F the *first* terms, O the *outer* terms, I the *inner* terms, and L the *last* terms.

Example 1 Find $4y(6 - 2y + 5y^2)$.

$$4y(6 - 2y + 5y^2) = 4y(6) + 4y(-2y) + 4y(5y^2)$$ Distributive Property
$$= 24y - 8y^2 + 20y^3$$ Multiply the monomials.

Example 2 Find $(6x - 5)(2x + 1)$.

$$(6x - 5)(2x + 1) = \underset{\text{First terms}}{6x \cdot 2x} + \underset{\text{Outer terms}}{6x \cdot 1} + \underset{\text{Inner terms}}{(-5) \cdot 2x} + \underset{\text{Last terms}}{(-5) \cdot 1}$$

$$= 12x^2 + 6x - 10x - 5$$ Multiply monomials.
$$= 12x^2 - 4x - 5$$ Add like terms.

Exercises

Find each product.

1. $2x(3x^2 - 5)$

2. $7a(6 - 2a - a^2)$

3. $-5y^2(y^2 + 2y - 3)$

4. $(x - 2)(x + 7)$

5. $(5 - 4x)(3 - 2x)$

6. $(2x - 1)(3x + 5)$

7. $(4x + 3)(x + 8)$

8. $(7x - 2)(2x - 7)$

9. $(3x - 2)(x + 10)$

10. $3(2a + 5c) - 2(4a - 6c)$

11. $2(a - 6)(2a + 7)$

12. $2x(x + 5) - x^2(3 - x)$

13. $(3t^2 - 8)(t^2 + 5)$

14. $(2r + 7)^2$

15. $(c + 7)(c - 3)$

16. $(5a + 7)(5a - 7)$

17. $(3x^2 - 1)(2x^2 + 5x)$

18. $(x^2 - 2)(x^2 - 5)$

19. $(x + 1)(2x^2 - 3x + 1)$

20. $(2n^2 - 3)(n^2 + 5n - 1)$

21. $(x - 1)(x^2 - 3x + 4)$

6-3 Study Guide and Intervention

Dividing Polynomials

Use Long Division To divide a polynomial by a monomial, use the properties of exponents from Lesson 6-1.

To divide a polynomial by a polynomial, use a long division pattern. Remember that only like terms can be added or subtracted.

Example 1 Simplify $\dfrac{12p^3t^2r - 21p^2qtr^2 - 9p^3tr}{3p^2tr}$.

$$\dfrac{12p^3t^2r - 21p^2qtr^2 - 9p^3tr}{3p^2tr} = \dfrac{12p^3t^2r}{3p^2tr} - \dfrac{21p^2qtr^2}{3p^2tr} - \dfrac{9p^3tr}{3p^2tr}$$

$$= \dfrac{12}{3}p^{3-2}t^{2-1}r^{1-1} - \dfrac{21}{3}p^{2-2}qt^{1-1}r^{2-1} - \dfrac{9}{3}p^{3-2}t^{1-1}r^{1-1}$$

$$= 4pt - 7qr - 3p$$

Example 2 Use long division to find $(x^3 - 8x^2 + 4x - 9) \div (x - 4)$.

$$
\begin{array}{r}
x^2 - 4x - 12 \\
x - 4 \overline{)\, x^3 - 8x^2 + 4x - 9} \\
\underline{(-)x^3 - 4x^2} \\
-4x^2 + 4x \\
\underline{(-)-4x^2 + 16x} \\
-12x - 9 \\
\underline{(-)-12x + 48} \\
-57
\end{array}
$$

The quotient is $x^2 - 4x - 12$, and the remainder is -57.

Therefore $\dfrac{x^3 - 8x^2 + 4x - 9}{x - 4} = x^2 - 4x - 12 - \dfrac{57}{x - 4}$.

Exercises

Simplify.

1. $\dfrac{18a^3 + 30a^2}{3a}$

2. $\dfrac{24mn^6 - 40m^2n^3}{4m^2n^3}$

3. $\dfrac{60a^2b^3 - 48b^4 + 84a^5b^2}{12ab^2}$

4. $(2x^2 - 5x - 3) \div (x - 3)$

5. $(m^2 - 3m - 7) \div (m + 2)$

6. $(p^3 - 6) \div (p - 1)$

7. $(t^3 - 6t^2 + 1) \div (t + 2)$

8. $(x^5 - 1) \div (x - 1)$

9. $(2x^3 - 5x^2 + 4x - 4) \div (x - 2)$

6-3 Study Guide and Intervention (continued)

Dividing Polynomials

Use Synthetic Division

Synthetic division	a procedure to divide a polynomial by a binomial using coefficients of the dividend and the value of r in the divisor $x - r$

Use synthetic division to find $(2x^3 - 5x^2 + 5x - 2) \div (x - 1)$.

Step 1	Write the terms of the dividend so that the degrees of the terms are in descending order. Then write just the coefficients.	$2x^3 - 5x^2 + 5x - 2$ 2　−5　　5　−2
Step 2	Write the constant r of the divisor $x - r$ to the left, In this case, $r = 1$. Bring down the first coefficient, 2, as shown.	1⌋ 2　−5　　5　−2 ‾‾‾‾‾‾‾‾‾‾‾‾ 2
Step 3	Multiply the first coefficient by r, $1 \cdot 2 = 2$. Write their product under the second coefficient. Then add the product and the second coefficient: $-5 + 2 = -3$.	1⌋ 2　−5　　5　−2 　　　2 ‾‾‾‾‾‾‾‾‾‾‾‾ 2　−3
Step 4	Multiply the sum, -3, by r: $-3 \cdot 1 = -3$. Write the product under the next coefficient and add: $5 + (-3) = 2$.	1⌋ 2　−5　　5　−2 　　　2　−3 ‾‾‾‾‾‾‾‾‾‾‾‾ 2　−3　　2
Step 5	Multiply the sum, 2, by r: $2 \cdot 1 = 2$. Write the product under the next coefficient and add: $-2 + 2 = 0$. The remainder is 0.	1⌋ 2　−5　　5　−2 　　　2　−3　　2 ‾‾‾‾‾‾‾‾‾‾‾‾ 2　−3　　2　　0

Thus, $(2x^3 - 5x^2 + 5x - 2) \div (x - 1) = 2x^2 - 3x + 2$.

Exercises

Simplify.

1. $(3x^3 - 7x^2 + 9x - 14) \div (x - 2)$

2. $(5x^3 + 7x^2 - x - 3) \div (x + 1)$

3. $(2x^3 + 3x^2 - 10x - 3) \div (x + 3)$

4. $(x^3 - 8x^2 + 19x - 9) \div (x - 4)$

5. $(2x^3 + 10x^2 + 9x + 38) \div (x + 5)$

6. $(3x^3 - 8x^2 + 16x - 1) \div (x - 1)$

7. $(x^3 - 9x^2 + 17x - 1) \div (x - 2)$

8. $(4x^3 - 25x^2 + 4x + 20) \div (x - 6)$

9. $(6x^3 + 28x^2 - 7x + 9) \div (x + 5)$

10. $(x^4 - 4x^3 + x^2 + 7x - 2) \div (x - 2)$

11. $(12x^4 + 20x^3 - 24x^2 + 20x + 35) \div (3x + 5)$

6-4 Study Guide and Intervention
Polynomial Functions

Polynomial Functions

Polynomial in One Variable	A polynomial of degree n in one variable x is an expression of the form $a_0 x^n + a_1 x^{n-1} + \ldots + a_{n-2} x^2 + a_{n-1} x + a_n,$ where the coefficients $a_0, a_1, a_2, \ldots, a_n$ represent real numbers, a_0 is not zero, and n represents a nonnegative integer.

The **degree of a polynomial** in one variable is the greatest exponent of its variable. The **leading coefficient** is the coefficient of the term with the highest degree.

Polynomial Function	A polynomial function of degree n can be described by an equation of the form $P(x) = a_0 x^n + a_1 x^{n-1} + \ldots + a_{n-2} x^2 + a_{n-1} x + a_n,$ where the coefficients $a_0, a_1, a_2, \ldots, a_n$ represent real numbers, a_0 is not zero, and n represents a nonnegative integer.

Example 1 **What are the degree and leading coefficient of $3x^2 - 2x^4 - 7 + x^3$?**

Rewrite the expression so the powers of x are in decreasing order.

$-2x^4 + x^3 + 3x^2 - 7$

This is a polynomial in one variable. The degree is 4, and the leading coefficient is -2.

Example 2 **Find $f(-5)$ if $f(x) = x^3 + 2x^2 - 10x + 20$.**

$\begin{aligned} f(x) &= x^3 + 2x^2 - 10x + 20 && \text{Original function} \\ f(-5) &= (-5)^3 + 2(-5)^2 - 10(-5) + 20 && \text{Replace } x \text{ with } -5. \\ &= -125 + 50 + 50 + 20 && \text{Evaluate.} \\ &= -5 && \text{Simplify.} \end{aligned}$

Example 3 **Find $g(a^2 - 1)$ if $g(x) = x^2 + 3x - 4$.**

$\begin{aligned} g(x) &= x^2 + 3x - 4 && \text{Original function} \\ g(a^2 - 1) &= (a^2 - 1)^2 + 3(a^2 - 1) - 4 && \text{Replace } x \text{ with } a^2 - 1. \\ &= a^4 - 2a^2 + 1 + 3a^2 - 3 - 4 && \text{Evaluate.} \\ &= a^4 + a^2 - 6 && \text{Simplify.} \end{aligned}$

Exercises

State the degree and leading coefficient of each polynomial in one variable. If it is not a polynomial in one variable, explain why.

1. $3x^4 + 6x^3 - x^2 + 12$

2. $100 - 5x^3 + 10x^7$

3. $4x^6 + 6x^4 + 8x^8 - 10x^2 + 20$

4. $4x^2 - 3xy + 16y^2$

5. $8x^3 - 9x^5 + 4x^2 - 36$

6. $\dfrac{x^2}{18} - \dfrac{x^6}{25} + \dfrac{x^3}{36} - \dfrac{1}{72}$

Find $f(2)$ and $f(-5)$ for each function.

7. $f(x) = x^2 - 9$

8. $f(x) = 4x^3 - 3x^2 + 2x - 1$

9. $f(x) = 9x^3 - 4x^2 + 5x + 7$

6-4 Study Guide and Intervention (continued)

Polynomial Functions

Graphs of Polynomial Functions

End Behavior of Polynomial Functions	If the degree is even and the leading coefficient is positive, then
	$f(x) \to +\infty$ as $x \to -\infty$
	$f(x) \to +\infty$ as $x \to +\infty$
	If the degree is even and the leading coefficient is negative, then
	$f(x) \to -\infty$ as $x \to -\infty$
	$f(x) \to -\infty$ as $x \to +\infty$
	If the degree is odd and the leading coefficient is positive, then
	$f(x) \to -\infty$ as $x \to -\infty$
	$f(x) \to +\infty$ as $x \to +\infty$
	If the degree is odd and the leading coefficient is negative, then
	$f(x) \to +\infty$ as $x \to -\infty$
	$f(x) \to -\infty$ as $x \to +\infty$

Real Zeros of a Polynomial Function	The maximum number of zeros of a polynomial function is equal to the degree of the polynomial. A zero of a function is a point at which the graph intersects the x-axis. On a graph, count the number of real zeros of the function by counting the number of times the graph crosses or touches the x-axis.

Example Determine whether the graph represents an odd-degree polynomial or an even-degree polynomial. Then state the number of real zeros.

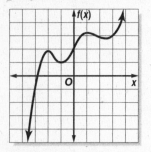

As $x \to -\infty$, $f(x) \to -\infty$ and as $x \to +\infty$, $f(x) \to +\infty$, so it is an odd-degree polynomial function.
The graph intersects the x-axis at 1 point, so the function has 1 real zero.

Exercises

Determine whether each graph represents an odd-degree polynomial or an even-degree polynomial. Then state the number of real zeros.

1.
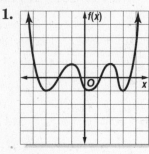

2.
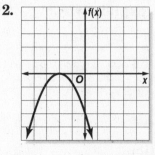

3.
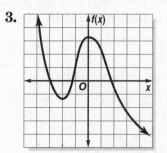

6-5 Study Guide and Intervention

Analyze Graphs of Polynomial Functions

Graph Polynomial Functions

Location Principle	Suppose $y = f(x)$ represents a polynomial function and a and b are two numbers such that $f(a) < 0$ and $f(b) > 0$. Then the function has at least one real zero between a and b.

Example Determine the values of x between which each real zero of the function $f(x) = 2x^4 - x^3 - 5$ is located. Then draw the graph.

Make a table of values. Look at the values of $f(x)$ to locate the zeros. Then use the points to sketch a graph of the function.

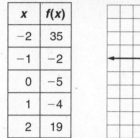

x	$f(x)$
-2	35
-1	-2
0	-5
1	-4
2	19

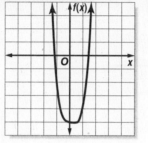

The changes in sign indicate that there are zeros between $x = -2$ and $x = -1$ and between $x = 1$ and $x = 2$.

Exercises

Graph each function by making a table of values. Determine the values of x at which or between which each real zero is located.

1. $f(x) = x^3 - 2x^2 + 1$

2. $f(x) = x^4 + 2x^3 - 5$

3. $f(x) = -x^4 + 2x^2 - 1$

4. $f(x) = x^3 - 3x^2 + 4$

5. $f(x) = 3x^3 + 2x - 1$

6. $f(x) = x^4 - 3x^3 + 1$

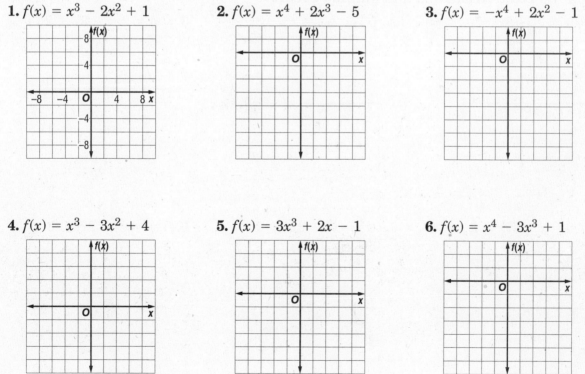

6-5 Study Guide and Intervention (continued)

Analyze Graphs of Polynomial Functions

Maximum and Minimum Points A quadratic function has either a maximum or a minimum point on its graph. For higher degree polynomial functions, you can find *turning points*, which represent **relative maximum** or **relative minimum** points.

Example Graph $f(x) = x^3 + 6x^2 - 3$. Estimate the *x*-coordinates at which the relative maxima and minima occur.

Make a table of values and graph the function.

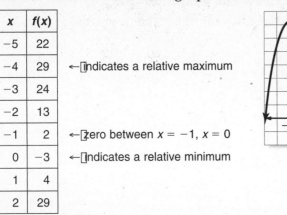

x	f(x)	
−5	22	
−4	29	←indicates a relative maximum
−3	24	
−2	13	
−1	2	←zero between $x = -1$, $x = 0$
0	−3	←indicates a relative minimum
1	4	
2	29	

A relative maximum occurs at $x = -4$ and a relative minimum occurs at $x = 0$.

Exercises

Graph each function by making a table of values. Estimate the *x*-coordinates at which the relative maxima and minima occur.

1. $f(x) = x^3 - 3x^2$

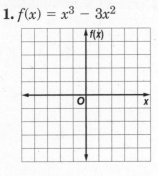

2. $f(x) = 2x^3 + x^2 - 3x$

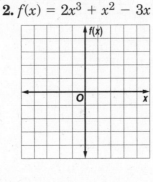

3. $f(x) = 2x^3 - 3x + 2$

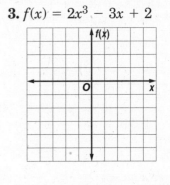

4. $f(x) = x^4 - 7x - 3$

5. $f(x) = x^5 - 2x^2 + 2$

6. $f(x) = x^3 + 2x^2 - 3$

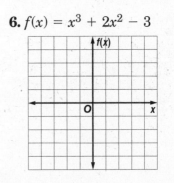

6-6 Study Guide and Intervention

Solving Polynomial Equations

Factor Polynomials

Techniques for Factoring Polynomials	For any number of terms, check for: greatest common factor
	For two terms, check for: Difference of two squares $a^2 - b^2 = (a + b)(a - b)$ Sum of two cubes $a^3 + b^3 = (a + b)(a^2 - ab + b^2)$ Difference of two cubes $a^3 - b^3 = (a - b)(a^2 + ab + b^2)$
	For three terms, check for: Perfect square trinomials $a^2 + 2ab + b^2 = (a + b)^2$ $a^2 - 2ab + b^2 = (a - b)^2$ General trinomials $acx^2 + (ad + bc)x + bd = (ax + b)(cx + d)$
	For four terms, check for: Grouping $ax + bx + ay + by = x(a + b) + y(a + b)$ $= (a + b)(x + y)$

Example   Factor $24x^2 - 42x - 45$.

First factor out the GCF to get $24x^2 - 42x - 45 = 3(8x^2 - 14x - 15)$. To find the coefficients of the x terms, you must find two numbers whose product is $8 \cdot (-15) = -120$ and whose sum is -14. The two coefficients must be -20 and 6. Rewrite the expression using $-20x$ and $6x$ and factor by grouping.

$$8x^2 - 14x - 15 = 8x^2 - 20x + 6x - 15 \qquad \text{Group to find a GCF.}$$
$$= 4x(2x - 5) + 3(2x - 5) \qquad \text{Factor the GCF of each binomial.}$$
$$= (4x + 3)(2x - 5) \qquad \text{Distributive Property}$$

Thus, $24x^2 - 42x - 45 = 3(4x + 3)(2x - 5)$.

Exercises

Factor completely. If the polynomial is not factorable, write *prime*.

1. $14x^2y^2 + 42xy^3$ **2.** $6mn + 18m - n - 3$ **3.** $2x^2 + 18x + 16$

4. $x^4 - 1$ **5.** $35x^3y^4 - 60x^4y$ **6.** $2r^3 + 250$

7. $100m^8 - 9$ **8.** $x^2 + x + 1$ **9.** $c^4 + c^3 - c^2 - c$

6-6 Study Guide and Intervention (continued)

Solving Polynomial Equations

Solve Equations Using Quadratic Form If a polynomial expression can be written in quadratic form, then you can use what you know about solving quadratic equations to solve the related polynomial equation.

Example 1 Solve $x^4 - 40x^2 + 144 = 0$.

$x^4 - 40x^2 + 144 = 0$ Original equation

$(x^2)^2 - 40(x^2) + 144 = 0$ Write the expression on the left in quadratic form.

$(x^2 - 4)(x^2 - 36) = 0$ Factor.

$x^2 - 4 = 0$ or $x^2 - 36 = 0$ Zero Product Property

$(x - 2)(x + 2) = 0$ or $(x - 6)(x + 6) = 0$ Factor.

$x - 2 = 0$ or $x + 2 = 0$ or $x - 6 = 0$ or $x + 6 = 0$ Zero Product Property

$x = 2$ or $x = -2$ or $x = 6$ or $x = -6$ Simplify.

The solutions are ± 2 and ± 6.

Example 2 Solve $2x + \sqrt{x} - 15 = 0$.

$2x + \sqrt{x} - 15 = 0$ Original equation

$2(\sqrt{x})^2 + \sqrt{x} - 15 = 0$ Write the expression on the left in quadratic form.

$(2\sqrt{x} - 5)(\sqrt{x} + 3) = 0$ Factor.

$2\sqrt{x} - 5 = 0$ or $\sqrt{x} + 3 = 0$ Zero Product Property

$\sqrt{x} = \dfrac{5}{2}$ or $\sqrt{x} = -3$ Simplify.

Since the principal square root of a number cannot be negative, $\sqrt{x} = -3$ has no solution. The solution is $\dfrac{25}{4}$ or $6\dfrac{1}{4}$.

Exercises

Solve each equation.

1. $x^4 = 49$

2. $x^4 - 6x^2 = -8$

3. $x^4 - 3x^2 = 54$

4. $3t^6 - 48t^2 = 0$

5. $m^6 - 16m^3 + 64 = 0$

6. $y^4 - 5y^2 + 4 = 0$

7. $x^4 - 29x^2 + 100 = 0$

8. $4x^4 - 73x^2 + 144 = 0$

9. $\dfrac{1}{x^2} - \dfrac{7}{x} + 12 = 0$

10. $x - 5\sqrt{x} + 6 = 0$

11. $x - 10\sqrt{x} + 21 = 0$

12. $x^{\frac{2}{3}} - 5x^{\frac{1}{3}} + 6 = 0$

6-7 Study Guide and Intervention

The Remainder and Factor Theorems

Synthetic Substitution

Remainder Theorem	The remainder, when you divide the polynomial $f(x)$ by $(x - a)$, is the constant $f(a)$. $f(x) = q(x) \cdot (x - a) + f(a)$, where $q(x)$ is a polynomial with degree one less than the degree of $f(x)$.

Example 1   If $f(x) = 3x^4 + 2x^3 - 5x^2 + x - 2$, find $f(-2)$.

Method 1 Synthetic Substitution

By the Remainder Theorem, $f(-2)$ should be the remainder when you divide the polynomial by $x + 2$.

```
-2 |  3   2  -5   1  -2
   |     -6   8  -6  10
     3  -4   3  -5 |  8
```

The remainder is 8, so $f(-2) = 8$.

Method 2 Direct Substitution

Replace x with -2.

$f(x) = 3x^4 + 2x^3 - 5x^2 + x - 2$
$f(-2) = 3(-2)^4 + 2(-2)^3 - 5(-2)^2 + (-2) - 2$
$= 48 - 16 - 20 - 2 - 2$ or 8

So $f(-2) = 8$.

Example 2   If $f(x) = 5x^3 + 2x - 1$, find $f(3)$.

Again, by the Remainder Theorem, $f(3)$ should be the remainder when you divide the polynomial by $x - 3$.

```
3 |  5   0    2   -1
  |     15   45  141
    5   15   47 | 140
```

The remainder is 140, so $f(3) = 140$.

Exercises

Use synthetic substitution to find $f(-5)$ and $f\left(\dfrac{1}{2}\right)$ for each function.

1. $f(x) = -3x^2 + 5x - 1$

2. $f(x) = 4x^2 + 6x - 7$

3. $f(x) = -x^3 + 3x^2 - 5$

4. $f(x) = x^4 + 11x^2 - 1$

Use synthetic substitution to find $f(4)$ and $f(-3)$ for each function.

5. $f(x) = 2x^3 + x^2 - 5x + 3$

6. $f(x) = 3x^3 - 4x + 2$

7. $f(x) = 5x^3 - 4x^2 + 2$

8. $f(x) = 2x^4 - 4x^3 + 3x^2 + x - 6$

9. $f(x) = 5x^4 + 3x^3 - 4x^2 - 2x + 4$

10. $f(x) = 3x^4 - 2x^3 - x^2 + 2x - 5$

11. $f(x) = 2x^4 - 4x^3 - x^2 - 6x + 3$

12. $f(x) = 4x^4 - 4x^3 + 3x^2 - 2x - 3$

6-7 Study Guide and Intervention (continued)

The Remainder and Factor Theorems

Factors of Polynomials The **Factor Theorem** can help you find all the factors of a polynomial.

Factor Theorem	The binomial $x - a$ is a factor of the polynomial $f(x)$ if and only if $f(a) = 0$.

Example Show that $x + 5$ is a factor of $x^3 + 2x^2 - 13x + 10$. Then find the remaining factors of the polynomial.

By the Factor Theorem, the binomial $x + 5$ is a factor of the polynomial if -5 is a zero of the polynomial function. To check this, use synthetic substitution.

$$
\begin{array}{r|rrrr}
-5 & 1 & 2 & -13 & 10 \\
   &   & -5 & 15 & -10 \\
\hline
   & 1 & -3 & 2 & 0 \\
\end{array}
$$

Since the remainder is 0, $x + 5$ is a factor of the polynomial. The polynomial $x^3 + 2x^2 - 13x + 10$ can be factored as $(x + 5)(x^2 - 3x + 2)$. The depressed polynomial $x^2 - 3x + 2$ can be factored as $(x - 2)(x - 1)$.

So $x^3 + 2x^2 - 13x + 10 = (x + 5)(x - 2)(x - 1)$.

Exercises

Given a polynomial and one of its factors, find the remaining factors of the polynomial. Some factors may not be binomials.

1. $x^3 + x^2 - 10x + 8; x - 2$

2. $x^3 - 4x^2 - 11x + 30; x + 3$

3. $x^3 + 15x^2 + 71x + 105; x + 7$

4. $x^3 - 7x^2 - 26x + 72; x + 4$

5. $2x^3 - x^2 - 7x + 6; x - 1$

6. $3x^3 - x^2 - 62x - 40; x + 4$

7. $12x^3 - 71x^2 + 57x - 10; x - 5$

8. $14x^3 + x^2 - 24x + 9; x - 1$

9. $x^3 + x + 10; x + 2$

10. $2x^3 - 11x^2 + 19x - 28; x - 4$

11. $3x^3 - 13x^2 - 34x + 24; x - 6$

12. $x^4 + x^3 - 11x^2 - 9x + 18; x - 1$

6-8 Study Guide and Intervention

Roots and Zeros

Types of Roots The following statements are equivalent for any polynomial function $f(x)$.

- c is a zero of the polynomial function $f(x)$.
- $(x - c)$ is a factor of the polynomial $f(x)$.
- c is a root or solution of the polynomial equation $f(x) = 0$.

If c is real, then $(c, 0)$ is an intercept of the graph of $f(x)$.

Fundamental Theorem of Algebra	Every polynomial equation with degree greater than zero has at least one root in the set of complex numbers.
Corollary to the Fundamental Theorem of Algebras	A polynomial equation of the form $P(x) = 0$ of degree n with complex coefficients has exactly n roots in the set of complex numbers.
Descartes' Rule of Signs	If $P(x)$ is a polynomial with real coefficients whose terms are arranged in descending powers of the variable, • the number of positive real zeros of $y = P(x)$ is the same as the number of changes in sign of the coefficients of the terms, or is less than this by an even number, and • the number of negative real zeros of $y = P(x)$ is the same as the number of changes in sign of the coefficients of the terms of $P(-x)$, or is less than this number by an even number.

Example 1 Solve the equation $6x^3 + 3x = 0$ and state the number and type of roots.

$$6x^3 + 3x = 0$$
$$3x(2x^2 + 1) = 0$$

Use the Zero Product Property.

$$3x = 0 \text{ or } 2x^2 + 1 = 0$$
$$x = 0 \text{ or } \qquad 2x^2 = -1$$
$$x = \pm \frac{i\sqrt{2}}{2}$$

The equation has one real root, 0, and two imaginary roots, $\pm \dfrac{i\sqrt{2}}{2}$.

Example 2 State the number of positive real zeros, negative real zeros, and imaginary zeros for $p(x) = 4x^4 - 3x^3 + x^2 + 2x - 5$.

Since $p(x)$ has degree 4, it has 4 zeros.

Since there are three sign changes, there are 3 or 1 positive real zeros.

Find $p(-x)$ and count the number of changes in sign for its coefficients.

$$p(-x) = 4(-x)^4 - 3(-x)^3 + (-x)^2 + 2(-x) - 5$$
$$= 4x^4 + 3x^3 + x^2 - 2x - 5$$

Since there is one sign change, there is exactly 1 negative real zero.

Thus, there are 3 positive and 1 negative real zero or 1 positive and 1 negative real zeros and 2 imaginary zeros.

Exercises

Solve each equation and state the number and type of roots.

1. $x^2 + 4x - 21 = 0$ **2.** $2x^3 - 50x = 0$ **3.** $12x^3 + 100x = 0$

State the number of positive real zeros, negative real zeros, and imaginary zeros for each function.

4. $f(x) = 3x^3 + x^2 - 8x - 12$ **5.** $f(x) = 3x^5 - x^4 - x^3 + 6x^2 - 5$

6-8 Study Guide and Intervention (continued)

Roots and Zeros

Find Zeros

Complex Conjugate Theorem	Suppose a and b are real numbers with $b \neq 0$. If $a + bi$ is a zero of a polynomial function with real coefficients, then $a - bi$ is also a zero of the function.

Example   Find all of the zeros of $f(x) = x^4 - 15x^2 + 38x - 60$.

Since $f(x)$ has degree 4, the function has 4 zeros.

$f(x) = x^4 - 15x^2 + 38x - 60$ $f(-x) = x^4 - 15x^2 - 38x - 60$

Since there are 3 sign changes for the coefficients of $f(x)$, the function has 3 or 1 positive real zeros. Since there is 1 sign change for the coefficients of $f(-x)$, the function has 1 negative real zero. Use synthetic substitution to test some possible zeros.

$$
\begin{array}{r|rrrrr}
2 & 1 & 0 & -15 & 38 & -60 \\
  &   & 2 & 4 & -22 & 32 \\
\hline
  & 1 & 2 & -11 & 16 & -28 \\
\end{array}
$$

$$
\begin{array}{r|rrrrr}
3 & 1 & 0 & -15 & 38 & -60 \\
  &   & 3 & 9 & -18 & 60 \\
\hline
  & 1 & 3 & -6 & 20 & 0 \\
\end{array}
$$

So 3 is a zero of the polynomial function. Now try synthetic substitution again to find a zero of the depressed polynomial.

$$
\begin{array}{r|rrrr}
-2 & 1 & 3 & -6 & 20 \\
   &   & -2 & -2 & 16 \\
\hline
   & 1 & 1 & -8 & 36 \\
\end{array}
$$

$$
\begin{array}{r|rrrr}
-4 & 1 & 3 & -6 & 20 \\
   &   & -4 & 4 & 8 \\
\hline
   & 1 & -1 & -2 & 28 \\
\end{array}
$$

$$
\begin{array}{r|rrrr}
-5 & 1 & 3 & -6 & 20 \\
   &   & -5 & 10 & -20 \\
\hline
   & 1 & -2 & 4 & 0 \\
\end{array}
$$

So -5 is another zero. Use the Quadratic Formula on the depressed polynomial $x^2 - 2x + 4$ to find the other 2 zeros, $1 \pm i\sqrt{3}$.

The function has two real zeros at 3 and -5 and two imaginary zeros at $1 \pm i\sqrt{3}$.

Exercises

Find all of the zeros of each function.

1. $f(x) = x^3 + x^2 + 9x + 9$

2. $f(x) = x^3 - 3x^2 + 4x - 12$

3. $p(a) = a^3 - 10a^2 + 34a - 40$

4. $p(x) = x^3 - 5x^2 + 11x - 15$

5. $f(x) = x^3 + 6x + 20$

6. $f(x) = x^4 - 3x^3 + 21x^2 - 75x - 100$

6-9 Study Guide and Intervention

Rational Zero Theorem

Identify Rational Zeros

Rational Zero Theorem	Let $f(x) = a_0x^n + a_1x^{n-1} + \dots + a_{n-2}x^2 + a_{n-1}x + a^n$ represent a polynomial function with integral coefficients. If $\frac{p}{q}$ is a rational number in simplest form and is a zero of $y = f(x)$, then p is a factor of a_n and q is a factor of a_0.
Corollary (Integral Zero Theorem)	If the coefficients of a polynomial are integers such that $a_0 = 1$ and $a_n \neq 0$, any rational zeros of the function must be factors of a_n.

Example List all of the possible rational zeros of each function.

a. $f(x) = 3x^4 - 2x^2 + 6x - 10$

If $\frac{p}{q}$ is a rational root, then p is a factor of -10 and q is a factor of 3. The possible values for p are $\pm 1, \pm 2, \pm 5,$ and ± 10. The possible values for q are ± 1 and ± 3. So all of the possible rational zeros are $\frac{p}{q} = \pm 1, \pm 2, \pm 5, \pm 10, \pm\frac{1}{3}, \pm\frac{2}{3}, \pm\frac{5}{3},$ and $\pm\frac{10}{3}$.

b. $q(x) = x^3 - 10x^2 + 14x - 36$

Since the coefficient of x^3 is 1, the possible rational zeros must be the factors of the constant term -36. So the possible rational zeros are $\pm 1, \pm 2, \pm 3, \pm 4, \pm 6, \pm 9, \pm 12, \pm 18,$ and ± 36.

Exercises

List all of the possible rational zeros of each function.

1. $f(x) = x^3 + 3x^2 - x + 8$

2. $g(x) = x^5 - 7x^4 + 3x^2 + x - 20$

3. $h(x) = x^4 - 7x^3 - 4x^2 + x - 49$

4. $p(x) = 2x^4 - 5x^3 + 8x^2 + 3x - 5$

5. $q(x) = 3x^4 - 5x^3 + 10x + 12$

6. $r(x) = 4x^5 - 2x + 18$

7. $f(x) = x^7 - 6x^5 - 3x^4 + x^3 + 4x^2 - 120$

8. $g(x) = 5x^6 - 3x^4 + 5x^3 + 2x^2 - 15$

9. $h(x) = 6x^5 - 3x^4 + 12x^3 + 18x^2 - 9x + 21$

10. $p(x) = 2x^7 - 3x^6 + 11x^5 - 20x^2 + 11$

6-9 Study Guide and Intervention (continued)

Rational Zero Theorem

Find Rational Zeros

Example 1 Find all of the rational zeros of $f(x) = 5x^3 + 12x^2 - 29x + 12$.

From the corollary to the Fundamental Theorem of Algebra, we know that there are exactly 3 complex roots. According to Descartes' Rule of Signs there are 2 or 0 positive real roots and 1 negative real root. The possible rational zeros are $\pm 1, \pm 2, \pm 3, \pm 4, \pm 6, \pm 12,$ $\pm \frac{1}{5}, \pm \frac{2}{5}, \pm \frac{3}{5}, \pm \frac{4}{5}, \pm \frac{6}{5}, \pm \frac{12}{5}$. Make a table and test some possible rational zeros.

$\frac{p}{q}$	5	12	−29	12
1	5	17	−12	0

Since $f(1) = 0$, you know that $x = 1$ is a zero.
The depressed polynomial is $5x^2 + 17x - 12$, which can be factored as $(5x - 3)(x + 4)$.
By the Zero Product Property, this expression equals 0 when $x = \frac{3}{5}$ or $x = -4$.
The rational zeros of this function are $1, \frac{3}{5}$, and -4.

Example 2 Find all of the zeros of $f(x) = 8x^4 + 2x^3 + 5x^2 + 2x - 3$.

There are 4 complex roots, with 1 positive real root and 3 or 1 negative real roots. The possible rational zeros are $\pm 1, \pm 3, \pm \frac{1}{2}, \pm \frac{1}{4}, \pm \frac{1}{8}, \pm \frac{3}{2}, \pm \frac{3}{4},$ and $\pm \frac{3}{8}$.

Make a table and test some possible values.

$\frac{p}{q}$	8	2	5	2	−3
1	8	10	15	17	14
2	8	18	41	84	165
$\frac{1}{2}$	8	6	8	6	0

Since $f\left(\frac{1}{2}\right) = 0$, we know that $x = \frac{1}{2}$ is a root.

The depressed polynomial is $8x^3 + 6x^2 + 8x + 6$.
Try synthetic substitution again. Any remaining rational roots must be negative.

$\frac{p}{q}$	8	6	8	6
$-\frac{1}{4}$	8	4	7	$4\frac{1}{4}$
$-\frac{3}{4}$	8	0	8	0

$x = -\frac{3}{4}$ is another rational root.
The depressed polynomial is $8x^2 + 8 = 0$, which has roots $\pm i$.

The zeros of this function are $\frac{1}{2}, -\frac{3}{4}$, and $\pm i$.

Exercises

Find all of the rational zeros of each function.

1. $f(x) = x^3 + 4x^2 - 25x - 28$

2. $f(x) = x^3 + 6x^2 + 4x + 24$

Find all of the zeros of each function.

3. $f(x) = x^4 + 2x^3 - 11x^2 + 8x - 60$

4. $f(x) = 4x^4 + 5x^3 + 30x^2 + 45x - 54$

7-1 Study Guide and Intervention

Operations on Functions

Arithmetic Operations

Operations with Functions	Sum	$(f + g)(x) = f(x) + g(x)$
	Difference	$(f - g)(x) = f(x) - g(x)$
	Product	$(f \cdot g)(x) = f(x) \cdot g(x)$
	Quotient	$\left(\dfrac{f}{g}\right)(x) = \dfrac{f(x)}{g(x)}, g(x) \neq 0$

Example Find $(f + g)(x)$, $(f - g)(x)$, $(f \cdot g)(x)$, and $\left(\dfrac{f}{g}\right)(x)$ for $f(x) = x^2 + 3x - 4$ and $g(x) = 3x - 2$.

$(f + g)(x) = f(x) + g(x)$ Addition of functions

$\quad\quad\quad = (x^2 + 3x - 4) + (3x - 2)$ $f(x) = x^2 + 3x - 4, g(x) = 3x - 2$

$\quad\quad\quad = x^2 + 6x - 6$ Simplify.

$(f - g)(x) = f(x) - g(x)$ Subtraction of functions

$\quad\quad\quad = (x^2 + 3x - 4) - (3x - 2)$ $f(x) = x^2 + 3x - 4, g(x) = 3x - 2$

$\quad\quad\quad = x^2 - 2$ Simplify.

$(f \cdot g)(x) = f(x) \cdot g(x)$ Multiplication of functions

$\quad\quad\quad = (x^2 + 3x - 4)(3x - 2)$ $f(x) = x^2 + 3x - 4, g(x) = 3x - 2$

$\quad\quad\quad = x^2(3x - 2) + 3x(3x - 2) - 4(3x - 2)$ Distributive Property

$\quad\quad\quad = 3x^3 - 2x^2 + 9x^2 - 6x - 12x + 8$ Distributive Property

$\quad\quad\quad = 3x^3 + 7x^2 - 18x + 8$ Simplify.

$\left(\dfrac{f}{g}\right)(x) = \dfrac{f(x)}{g(x)}$ Division of functions

$\quad\quad\quad = \dfrac{x^2 + 3x - 4}{3x - 2}, x \neq \dfrac{2}{3}$ $f(x) = x^2 + 3x - 4$ and $g(x) = 3x - 2$

Exercises

Find $(f + g)(x)$, $(f - g)(x)$, $(f \cdot g)(x)$, and $\left(\dfrac{f}{g}\right)(x)$ for each $f(x)$ and $g(x)$.

1. $f(x) = 8x - 3; g(x) = 4x + 5$ **2.** $f(x) = x^2 + x - 6; g(x) = x - 2$

3. $f(x) = 3x^2 - x + 5; g(x) = 2x - 3$ **4.** $f(x) = 2x - 1; g(x) = 3x^2 + 11x - 4$

5. $f(x) = x^2 - 1; g(x) = \dfrac{1}{x + 1}$

7-1 Study Guide and Intervention (continued)

Operations on Functions

Composition of Functions

Composition of Functions	Suppose f and g are functions such that the range of g is a subset of the domain of f. Then the composite function $f \circ g$ can be described by the equation $[f \circ g](x) = f[g(x)]$.

Example 1 For $f = \{(1, 2), (3, 3), (2, 4), (4, 1)\}$ and $g = \{(1, 3), (3, 4), (2, 2), (4, 1)\}$, find $f \circ g$ and $g \circ f$ if they exist.

$f[g(1)] = f(3) = 3 \qquad f[g(2)] = f(2) = 4 \qquad f[g(3)] = f(4) = 1 \qquad f[g(4)] = f(1) = 2$

$f \circ g = \{(1, 3), (2, 4), (3, 1), (4, 2)\}$

$g[f(1)] = g(2) = 2 \qquad g[f(2)] = g(4) = 1 \qquad g[f(3)] = g(3) = 4 \qquad g[f(4)] = g(1) = 3$

$g \circ f = \{(1, 2), (2, 1), (3, 4), (4, 3)\}$

Example 2 Find $[g \circ h](x)$ and $[h \circ g](x)$ for $g(x) = 3x - 4$ and $h(x) = x^2 - 1$.

$$\begin{aligned}
[g \circ h](x) &= g[h(x)] & [h \circ g](x) &= h[g(x)] \\
&= g(x^2 - 1) & &= h(3x - 4) \\
&= 3(x^2 - 1) - 4 & &= (3x - 4)^2 - 1 \\
&= 3x^2 - 7 & &= 9x^2 - 24x + 16 - 1 \\
& & &= 9x^2 - 24x + 15
\end{aligned}$$

Exercises

For each set of ordered pairs, find $f \circ g$ and $g \circ f$ if they exist.

1. $f = \{(-1, 2), (5, 6), (0, 9)\}$,
   $g = \{(6, 0), (2, -1), (9, 5)\}$

2. $f = \{(5, -2), (9, 8), (-4, 3), (0, 4)\}$,
   $g = \{(3, 7), (-2, 6), (4, -2), (8, 10)\}$

Find $[f \circ g](x)$ and $[g \circ f](x)$.

3. $f(x) = 2x + 7; g(x) = -5x - 1$

4. $f(x) = x^2 - 1; g(x) = -4x^2$

5. $f(x) = x^2 + 2x; g(x) = x - 9$

6. $f(x) = 5x + 4; g(x) = 3 - x$

88 *Glencoe Algebra 2*

7-2 Study Guide and Intervention

Inverse Functions and Relations

Find Inverses

Inverse Relations	Two relations are inverse relations if and only if whenever one relation contains the element (a, b), the other relation contains the element (b, a).
Property of Inverse Functions	Suppose f and f^{-1} are inverse functions. Then $f(a) = b$ if and only if $f^{-1}(b) = a$.

Example Find the inverse of the function $f(x) = \frac{2}{5}x - \frac{1}{5}$. Then graph the function and its inverse.

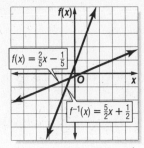

Step 1 Replace $f(x)$ with y in the original equation.

$$f(x) = \frac{2}{5}x - \frac{1}{5} \rightarrow y = \frac{2}{5}x - \frac{1}{5}$$

Step 2 Interchange x and y.

$$x = \frac{2}{5}y - \frac{1}{5}$$

Step 3 Solve for y.

$x = \frac{2}{5}y - \frac{1}{5}$ Inverse

$5x = 2y - 1$ Multiply each side by 5.

$5x + 1 = 2y$ Add 1 to each side.

$\frac{1}{2}(5x + 1) = y$ Divide each side by 2.

The inverse of $f(x) = \frac{2}{5}x - \frac{1}{5}$ is $f^{-1}(x) = \frac{1}{2}(5x + 1)$.

Exercises

Find the inverse of each function. Then graph the function and its inverse.

1. $f(x) = \frac{2}{3}x - 1$ **2.** $f(x) = 2x - 3$ **3.** $f(x) = \frac{1}{4}x - 2$

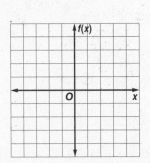   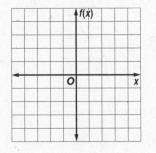   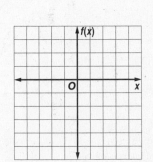

7-2   Study Guide and Intervention (continued)

Inverse Functions and Relations

Inverses of Relations and Functions

Inverse Functions	Two functions f and g are inverse functions if and only if $[f \circ g](x) = x$ and $[g \circ f](x) = x$.

Example 1   Determine whether $f(x) = 2x - 7$ and $g(x) = \frac{1}{2}(x + 7)$ are inverse functions.

$$[f \circ g](x) = f[g(x)] \qquad\qquad [g \circ f](x) = g[f(x)]$$
$$= f\left[\frac{1}{2}(x + 7)\right] \qquad\qquad = g(2x - 7)$$
$$= 2\left[\frac{1}{2}(x + 7)\right] - 7 \qquad\qquad = \frac{1}{2}(2x - 7 + 7)$$
$$= x + 7 - 7 \qquad\qquad = x$$
$$= x$$

The functions are inverses since both $[f \circ g](x) = x$ and $[g \circ f](x) = x$.

Example 2   Determine whether $f(x) = 4x + \frac{1}{3}$ and $g(x) = \frac{1}{4}x - 3$ are inverse functions.

$$[f \circ g](x) = f[g(x)]$$
$$= f\left(\frac{1}{4}x - 3\right)$$
$$= 4\left(\frac{1}{4}x - 3\right) + \frac{1}{3}$$
$$= x - 12 + \frac{1}{3}$$
$$= x - 11\frac{2}{3}$$

Since $[f \circ g](x) \neq x$, the functions are not inverses.

Exercises

Determine whether each pair of functions are inverse functions.

1. $f(x) = 3x - 1$
$\quad g(x) = \frac{1}{3}x + \frac{1}{3}$

2. $f(x) = \frac{1}{4}x + 5$
$\quad g(x) = 4x - 20$

3. $f(x) = \frac{1}{2}x - 10$
$\quad g(x) = 2x + \frac{1}{10}$

4. $f(x) = 2x + 5$
$\quad g(x) = 5x + 2$

5. $f(x) = 8x - 12$
$\quad g(x) = \frac{1}{8}x + 12$

6. $f(x) = -2x + 3$
$\quad g(x) = -\frac{1}{2}x + \frac{3}{2}$

7. $f(x) = 4x - \frac{1}{2}$
$\quad g(x) = \frac{1}{4}x + \frac{1}{8}$

8. $f(x) = 2x - \frac{3}{5}$
$\quad g(x) = \frac{1}{10}(5x + 3)$

9. $f(x) = 4x + \frac{1}{2}$
$\quad g(x) = \frac{1}{2}x - \frac{3}{2}$

10. $f(x) = 10 - \frac{x}{2}$
$\quad g(x) = 20 - 2x$

11. $f(x) = 4x - \frac{4}{5}$
$\quad g(x) = \frac{x}{4} + \frac{1}{5}$

12. $f(x) = 9 + \frac{3}{2}x$
$\quad g(x) = \frac{2}{3}x - 6$

7-3 Study Guide and Intervention
Square Root Functions and Inequalities

Square Root Functions A function that contains the square root of a variable expression is a **square root function**.

Example Graph $y = \sqrt{3x - 2}$. State its domain and range.

Since the radicand cannot be negative, $3x - 2 \geq 0$ or $x \geq \frac{2}{3}$.

The x-intercept is $\frac{2}{3}$. The range is $y \geq 0$.

Make a table of values and graph the function.

x	y
$\frac{2}{3}$	0
1	1
2	2
3	$\sqrt{7}$

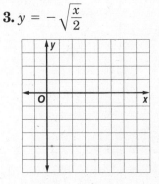

Exercises

Graph each function. State the domain and range of the function.

1. $y = \sqrt{2x}$

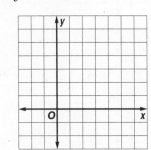

2. $y = -3\sqrt{x}$

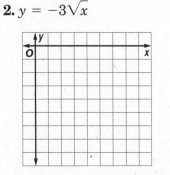

3. $y = -\sqrt{\dfrac{x}{2}}$

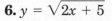

4. $y = 2\sqrt{x - 3}$

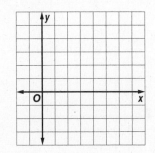

5. $y = -\sqrt{2x - 3}$

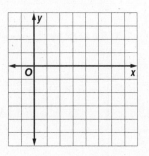

6. $y = \sqrt{2x + 5}$

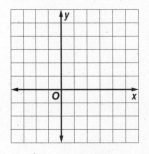

7-3 Study Guide and Intervention *(continued)*

Square Root Functions and Inequalities

Square Root Inequalities A **square root inequality** is an inequality that contains the square root of a variable expression. Use what you know about graphing square root functions and quadratic inequalities to graph square root inequalities.

Example   **Graph $y \leq \sqrt{2x - 1} + 2$.**

Graph the related equation $y = \sqrt{2x - 1} + 2$. Since the boundary should be included, the graph should be solid.

The domain includes values for $x \geq \dfrac{1}{2}$, so the graph is to the right of $x = \dfrac{1}{2}$.

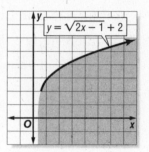

Exercises

Graph each inequality.

1. $y < 2\sqrt{x}$

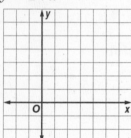

2. $y > \sqrt{x + 3}$

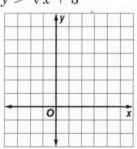

3. $y < 3\sqrt{2x - 1}$

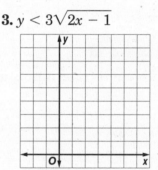

4. $y < \sqrt{3x - 4}$

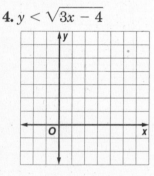

5. $y \geq \sqrt{x + 1} - 4$

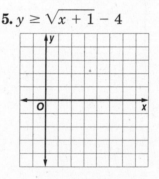

6. $y > 2\sqrt{2x - 3}$

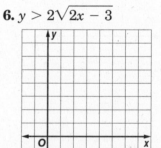

7. $y \geq \sqrt{3x + 1} - 2$

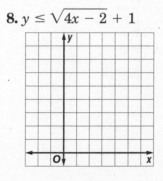

8. $y \leq \sqrt{4x - 2} + 1$

9. $y < 2\sqrt{2x - 1} - 4$

7-4 Study Guide and Intervention

nth Roots

Simplify Radicals

Square Root	For any real numbers a and b, if $a^2 = b$, then a is a square root of b.
nth Root	For any real numbers a and b, and any positive integer n, if $a^n = b$, then a is an nth root of b.
Real nth Roots of b, $\sqrt[n]{b}, -\sqrt[n]{b}$	1. If n is even and $b > 0$, then b has one positive root and one negative root. 2. If n is odd and $b > 0$, then b has one positive root. 3. If n is even and $b < 0$, then b has no real roots. 4. If n is odd and $b < 0$, then b has one negative root.

Example 1 Simplify $\sqrt{49z^8}$.

$\sqrt{49z^8} = \sqrt{(7z^4)^2} = 7z^4$

z^4 must be positive, so there is no need to take the absolute value.

Example 2 Simplify $-\sqrt[3]{(2a-1)^6}$

$-\sqrt[3]{(2a-1)^6} = -\sqrt[3]{[(2a-1)^2]^3} = (2a-1)^2$

Exercises

Simplify.

1. $\sqrt{81}$

2. $\sqrt[3]{-343}$

3. $\sqrt{144p^6}$

4. $\pm\sqrt{4a^{10}}$

5. $\sqrt[5]{243p^{10}}$

6. $-\sqrt[3]{m^6n^9}$

7. $\sqrt[3]{-b^{12}}$

8. $\sqrt{16a^{10}b^8}$

9. $\sqrt{121x^6}$

10. $\sqrt{(4k)^4}$

11. $\pm\sqrt{169r^4}$

12. $-\sqrt[3]{-27p^6}$

13. $-\sqrt{625y^2z^4}$

14. $\sqrt{36q^{34}}$

15. $\sqrt{100x^2y^4z^6}$

16. $\sqrt[3]{-0.027}$

17. $-\sqrt{-0.36}$

18. $\sqrt{0.64p^{10}}$

19. $\sqrt[4]{(2x)^8}$

20. $\sqrt{(11y^2)^4}$

21. $\sqrt[3]{(5a^2b)^6}$

22. $\sqrt{(3x-1)^2}$

23. $\sqrt[3]{(m-5)^6}$

24. $\sqrt{36x^2 - 12x + 1}$

7-4 Study Guide and Intervention (continued)

nth Roots

Approximate Radicals with a Calculator

Irrational Number	a number that cannot be expressed as a terminating or a repeating decimal

Radicals such as $\sqrt{2}$ and $\sqrt{3}$ are examples of irrational numbers. Decimal approximations for irrational numbers are often used in applications. These approximations can be easily found with a calculator.

Example Approximate $\sqrt[5]{18.2}$ with a calculator.

$\sqrt[5]{18.2} \approx 1.787$

Exercises

Use a calculator to approximate each value to three decimal places.

1. $\sqrt{62}$

2. $\sqrt{1050}$

3. $\sqrt[3]{0.054}$

4. $-\sqrt[4]{5.45}$

5. $\sqrt{5280}$

6. $\sqrt{18,600}$

7. $\sqrt{0.095}$

8. $\sqrt[3]{-15}$

9. $\sqrt[5]{100}$

10. $\sqrt[6]{856}$

11. $\sqrt{3200}$

12. $\sqrt{0.05}$

13. $\sqrt{12,500}$

14. $\sqrt{0.60}$

15. $-\sqrt[4]{500}$

16. $\sqrt[3]{0.15}$

17. $\sqrt[6]{4200}$

18. $\sqrt{75}$

19. **LAW ENFORCEMENT** The formula $r = 2\sqrt{5L}$ is used by police to estimate the speed r in miles per hour of a car if the length L of the car's skid mark is measures in feet. Estimate to the nearest tenth of a mile per hour the speed of a car that leaves a skid mark 300 feet long.

20. **SPACE TRAVEL** The distance to the horizon d miles from a satellite orbiting h miles above Earth can be approximated by $d = \sqrt{8000h + h^2}$. What is the distance to the horizon if a satellite is orbiting 150 miles above Earth?

7-5 Study Guide and Intervention

Operations with Radical Expressions

Simplify Radical Expressions

Product Property of Radicals	For any real numbers a and b, and any integer $n > 1$: 1. if n is even and a and b are both nonnegative, then $\sqrt[n]{ab} = \sqrt[n]{a} \cdot \sqrt[n]{b}$. 2. if n is odd, then $\sqrt[n]{ab} = \sqrt[n]{a} \cdot \sqrt[n]{b}$.

To simplify a square root, follow these steps:
1. Factor the radicand into as many squares as possible.
2. Use the Product Property to isolate the perfect squares.
3. Simplify each radical.

Quotient Property of Radicals	For any real numbers a and $b \neq 0$, and any integer $n > 1$, $\sqrt[n]{\dfrac{a}{b}} = \dfrac{\sqrt[n]{a}}{\sqrt[n]{b}}$, if all roots are defined.

To eliminate radicals from a denominator or fractions from a radicand, multiply the numerator and denominator by a quantity so that the radicand has an exact root.

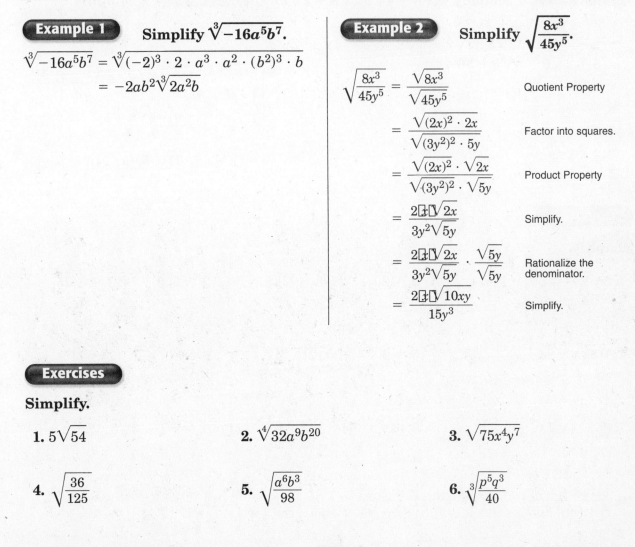

Example 1 Simplify $\sqrt[3]{-16a^5b^7}$.

$$\sqrt[3]{-16a^5b^7} = \sqrt[3]{(-2)^3 \cdot 2 \cdot a^3 \cdot a^2 \cdot (b^2)^3 \cdot b}$$
$$= -2ab^2\sqrt[3]{2a^2b}$$

Example 2 Simplify $\sqrt{\dfrac{8x^3}{45y^5}}$.

$$\sqrt{\dfrac{8x^3}{45y^5}} = \dfrac{\sqrt{8x^3}}{\sqrt{45y^5}} \qquad \text{Quotient Property}$$

$$= \dfrac{\sqrt{(2x)^2 \cdot 2x}}{\sqrt{(3y^2)^2 \cdot 5y}} \qquad \text{Factor into squares.}$$

$$= \dfrac{\sqrt{(2x)^2} \cdot \sqrt{2x}}{\sqrt{(3y^2)^2} \cdot \sqrt{5y}} \qquad \text{Product Property}$$

$$= \dfrac{2x\sqrt{2x}}{3y^2\sqrt{5y}} \qquad \text{Simplify.}$$

$$= \dfrac{2x\sqrt{2x}}{3y^2\sqrt{5y}} \cdot \dfrac{\sqrt{5y}}{\sqrt{5y}} \qquad \text{Rationalize the denominator.}$$

$$= \dfrac{2x\sqrt{10xy}}{15y^3} \qquad \text{Simplify.}$$

Exercises

Simplify.

1. $5\sqrt{54}$

2. $\sqrt[4]{32a^9b^{20}}$

3. $\sqrt{75x^4y^7}$

4. $\sqrt{\dfrac{36}{125}}$

5. $\sqrt{\dfrac{a^6b^3}{98}}$

6. $\sqrt[3]{\dfrac{p^5q^3}{40}}$

7-5 Study Guide and Intervention (continued)

Operations with Radical Expressions

Operations with Radicals When you add expressions containing radicals, you can add only like terms or **like radical expressions**. Two radical expressions are called *like radical expressions* if both the indices and the radicands are alike.

To multiply radicals, use the Product and Quotient Properties. For products of the form $(a\sqrt{b} + c\sqrt{d}) \cdot (e\sqrt{f} + g\sqrt{h})$, use the FOIL method. To rationalize denominators, use **conjugates**. Numbers of the form $a\sqrt{b} + c\sqrt{d}$ and $a\sqrt{b} - c\sqrt{d}$, where a, b, c, and d are rational numbers, are called conjugates. The product of conjugates is always a rational number.

Example 1 Simplify $2\sqrt{50} + 4\sqrt{500} - 6\sqrt{125}$.

$$
\begin{aligned}
2\sqrt{50} + 4\sqrt{500} - 6\sqrt{125} &= 2\sqrt{5^2 \cdot 2} + 4\sqrt{10^2 \cdot 5} - 6\sqrt{5^2 \cdot 5} \qquad \text{Factor using squares.}\\
&= 2 \cdot 5 \cdot \sqrt{2} + 4 \cdot 10 \cdot \sqrt{5} - 6 \cdot 5 \cdot \sqrt{5} \qquad \text{Simplify square roots.}\\
&= 10\sqrt{2} + 40\sqrt{5} - 30\sqrt{5} \qquad \text{Multiply.}\\
&= 10\sqrt{2} + 10\sqrt{5} \qquad \text{Combine like radicals.}
\end{aligned}
$$

Example 2 Simplify $(2\sqrt{3} - 4\sqrt{2})(\sqrt{3} + 2\sqrt{2})$.

$$
\begin{aligned}
&(2\sqrt{3} - 4\sqrt{2})(\sqrt{3} + 2\sqrt{2})\\
&= 2\sqrt{3} \cdot \sqrt{3} + 2\sqrt{3} \cdot 2\sqrt{2} - 4\sqrt{2} \cdot \sqrt{3} - 4\sqrt{2} \cdot 2\sqrt{2}\\
&= 6 + 4\sqrt{6} - 4\sqrt{6} - 16\\
&= -10
\end{aligned}
$$

Example 3 Simplify $\dfrac{2 - \sqrt{5}}{3 + \sqrt{5}}$.

$$
\begin{aligned}
\frac{2 - \sqrt{5}}{3 + \sqrt{5}} &= \frac{2 - \sqrt{5}}{3 + \sqrt{5}} \cdot \frac{3 - \sqrt{5}}{3 - \sqrt{5}}\\
&= \frac{6 - 2\sqrt{5} - 3\sqrt{5} + (\sqrt{5})^2}{3^2 - (\sqrt{5})^2}\\
&= \frac{6 - 5\sqrt{5} + 5}{9 - 5}\\
&= \frac{11 - 5\sqrt{5}}{4}
\end{aligned}
$$

Exercises

Simplify.

1. $3\sqrt{2} + \sqrt{50} - 4\sqrt{8}$

2. $\sqrt{20} + \sqrt{125} - \sqrt{45}$

3. $\sqrt{300} - \sqrt{27} - \sqrt{75}$

4. $\sqrt[3]{81} \cdot \sqrt[3]{24}$

5. $\sqrt[3]{2}(\sqrt[3]{4} + \sqrt[3]{12})$

6. $2\sqrt{3}(\sqrt{15} + \sqrt{60})$

7. $(2 + 3\sqrt{7})(4 + \sqrt{7})$

8. $(6\sqrt{3} - 4\sqrt{2})(3\sqrt{3} + \sqrt{2})$

9. $(4\sqrt{2} - 3\sqrt{5})(2\sqrt{20} + 5)$

10. $\dfrac{5\sqrt{48} + \sqrt{75}}{5\sqrt{3}}$

11. $\dfrac{4 + \sqrt{2}}{2 - \sqrt{2}}$

12. $\dfrac{5 - 3\sqrt{3}}{1 + 2\sqrt{3}}$

7-6 Study Guide and Intervention

Rational Exponents

Rational Exponents and Radicals

Definition of $b^{\frac{1}{n}}$	For any real number b and any positive integer n, $b^{\frac{1}{n}} = \sqrt[n]{b}$, except when $b < 0$ and n is even.
Definition of $b^{\frac{m}{n}}$	For any nonzero real number b, and any integers m and n, with $n > 1$, $b^{\frac{m}{n}} = \sqrt[n]{b^m} = (\sqrt[n]{b})^m$, except when $b < 0$ and n is even.

Example 1   Write $28^{\frac{1}{2}}$ in radical form.

Notice that $28 > 0$.

$$28^{\frac{1}{2}} = \sqrt{28}$$
$$= \sqrt{2^2 \cdot 7}$$
$$= \sqrt{2^2} \cdot \sqrt{7}$$
$$= 2\sqrt{7}$$

Example 2   Evaluate $\left(\dfrac{-8}{-125}\right)^{\frac{1}{3}}$.

Notice that $-8 < 0$, $-125 < 0$, and 3 is odd.

$$\left(\frac{-8}{-125}\right)^{\frac{1}{3}} = \frac{\sqrt[3]{-8}}{\sqrt[3]{-125}}$$
$$= \frac{-2}{-5}$$
$$= \frac{2}{5}$$

Exercises

Write each expression in radical form.

1. $11^{\frac{1}{7}}$ **2.** $15^{\frac{1}{3}}$ **3.** $300^{\frac{3}{2}}$

Write each radical using rational exponents.

4. $\sqrt{47}$ **5.** $\sqrt[3]{3a^5b^2}$ **6.** $\sqrt[4]{162p^5}$

Evaluate each expression.

7. $-27^{\frac{2}{3}}$ **8.** $\dfrac{5^{-\frac{1}{2}}}{2\sqrt{5}}$ **9.** $(0.0004)^{\frac{1}{2}}$

10. $8^{\frac{2}{3}} \cdot 4^{\frac{3}{2}}$ **11.** $\dfrac{144^{-\frac{1}{2}}}{27^{-\frac{1}{3}}}$ **12.** $\dfrac{16^{-\frac{1}{2}}}{(0.25)^{\frac{1}{2}}}$

7-6 Study Guide and Intervention (continued)

Rational Exponents

Simplify Expressions All the properties of powers from Lesson 6-1 apply to rational exponents. When you simplify expressions with rational exponents, leave the exponent in rational form, and write the expression with all positive exponents. Any exponents in the denominator must be positive integers.

When you simplify radical expressions, you may use rational exponents to simplify, but your answer should be in radical form. Use the smallest index possible.

Example 1 Simplify $y^{\frac{2}{3}} \cdot y^{\frac{3}{8}}$.

$$y^{\frac{2}{3}} \cdot y^{\frac{3}{8}} = y^{\frac{2}{3} + \frac{3}{8}} = y^{\frac{25}{24}}$$

Example 2 Simplify $\sqrt[4]{144x^6}$.

$$\sqrt[4]{144x^6} = (144x^6)^{\frac{1}{4}}$$
$$= (2^4 \cdot 3^2 \cdot x^6)^{\frac{1}{4}}$$
$$= (2^4)^{\frac{1}{4}} \cdot (3^2)^{\frac{1}{4}} \cdot (x^6)^{\frac{1}{4}}$$
$$= 2 \cdot 3^{\frac{1}{2}} \cdot x^{\frac{3}{2}} = 2x \cdot (3x)^{\frac{1}{2}} = 2x\sqrt{3x}$$

Exercises

Simplify each expression.

1. $x^{\frac{4}{5}} \cdot x^{\frac{6}{5}}$

2. $\left(y^{\frac{2}{3}}\right)^{\frac{3}{4}}$

3. $p^{\frac{4}{5}} \cdot p^{\frac{7}{10}}$

4. $\left(m^{-\frac{6}{5}}\right)^{\frac{2}{5}}$

5. $x^{-\frac{3}{8}} \cdot x^{\frac{4}{3}}$

6. $\left(s^{-\frac{1}{6}}\right)^{-\frac{4}{3}}$

7. $\dfrac{p}{p^{\frac{1}{3}}}$

8. $\left(a^{\frac{2}{3}}\right)^{\frac{6}{5}} \cdot \left(a^{\frac{2}{5}}\right)^{3}$

9. $\dfrac{x^{-\frac{1}{2}}}{x^{-\frac{1}{3}}}$

10. $\sqrt[6]{128}$

11. $\sqrt[4]{49}$

12. $\sqrt[5]{288}$

13. $\sqrt{32} \cdot 3\sqrt{16}$

14. $\sqrt[3]{25} \cdot \sqrt{125}$

15. $\sqrt[6]{16}$

16. $\dfrac{x - \sqrt[3]{3}}{\sqrt{12}}$

17. $\sqrt{\sqrt[3]{48}}$

18. $\dfrac{a\sqrt[3]{b^4}}{\sqrt{ab^3}}$

7-7 Study Guide and Intervention

Solving Radical Equations and Inequalities

Solve Radical Equations The following steps are used in solving equations that have variables in the radicand. Some algebraic procedures may be needed before you use these steps.

Step 1 Isolate the radical on one side of the equation.
Step 2 To eliminate the radical, raise each side of the equation to a power equal to the index of the radical.
Step 3 Solve the resulting equation.
Step 4 Check your solution in the original equation to make sure that you have not obtained any extraneous roots.

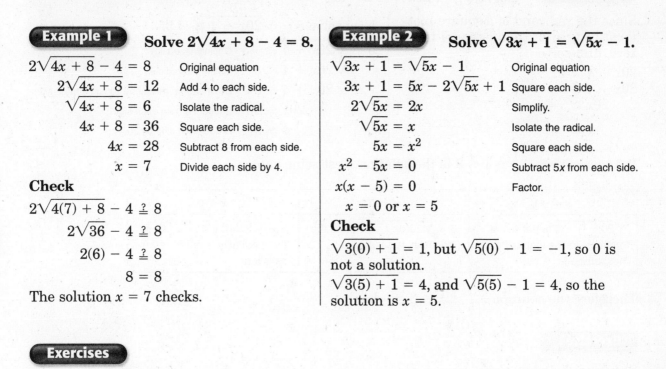

Example 1 Solve $2\sqrt{4x + 8} - 4 = 8$.

$2\sqrt{4x + 8} - 4 = 8$	Original equation
$2\sqrt{4x + 8} = 12$	Add 4 to each side.
$\sqrt{4x + 8} = 6$	Isolate the radical.
$4x + 8 = 36$	Square each side.
$4x = 28$	Subtract 8 from each side.
$x = 7$	Divide each side by 4.

Check

$2\sqrt{4(7) + 8} - 4 \stackrel{?}{=} 8$

$2\sqrt{36} - 4 \stackrel{?}{=} 8$

$2(6) - 4 \stackrel{?}{=} 8$

$8 = 8$

The solution $x = 7$ checks.

Example 2 Solve $\sqrt{3x + 1} = \sqrt{5x} - 1$.

$\sqrt{3x + 1} = \sqrt{5x} - 1$	Original equation
$3x + 1 = 5x - 2\sqrt{5x} + 1$	Square each side.
$2\sqrt{5x} = 2x$	Simplify.
$\sqrt{5x} = x$	Isolate the radical.
$5x = x^2$	Square each side.
$x^2 - 5x = 0$	Subtract 5x from each side.
$x(x - 5) = 0$	Factor.
$x = 0 \text{ or } x = 5$	

Check

$\sqrt{3(0) + 1} = 1$, but $\sqrt{5(0)} - 1 = -1$, so 0 is not a solution.

$\sqrt{3(5) + 1} = 4$, and $\sqrt{5(5)} - 1 = 4$, so the solution is $x = 5$.

Exercises

Solve each equation.

1. $3 + 2x\sqrt{3} = 5$

2. $2\sqrt{3x + 4} + 1 = 15$

3. $8 + \sqrt{x + 1} = 2$

4. $\sqrt{5 - x} - 4 = 6$

5. $12 + \sqrt{2x - 1} = 4$

6. $\sqrt{12 - x} = 0$

7. $\sqrt{21} - \sqrt{5x - 4} = 0$

8. $10 - \sqrt{2x} = 5$

9. $\sqrt{x^2 + 7x} = \sqrt{7x - 9}$

10. $4\sqrt[3]{2x + 11} - 2 = 10$

11. $2\sqrt{x + 11} = \sqrt{x + 2} + \sqrt{3x - 6}$

12. $\sqrt{9x - 11} = x + 1$

7-7 Study Guide and Intervention (continued)

Solving Radical Equations and Inequalities

Solve Radical Inequalities A **radical inequality** is an inequality that has a variable in a radicand. Use the following steps to solve radical inequalities.

Step 1	If the index of the root is even, identify the values of the variable for which the radicand is nonnegative.
Step 2	Solve the inequality algebraically.
Step 3	Test values to check your solution.

Example   Solve $5 - \sqrt{20x + 4} \geq -3.$

Since the radicand of a square root must be greater than or equal to zero, first solve

$20x + 4 \geq 0.$

$20x + 4 \geq 0$

$\quad 20x \geq -4$

$\qquad x \geq -\dfrac{1}{5}$

Now solve $5 - \sqrt{20x + 4} \geq -3.$

$5 - \sqrt{20x + 4} \geq -3$	Original inequality
$\sqrt{20x + 4} \leq 8$	Isolate the radical.
$20x + 4 \leq 64$	Eliminate the radical by squaring each side.
$20x \leq 60$	Subtract 4 from each side.
$x \leq 3$	Divide each side by 20.

It appears that $-\dfrac{1}{5} \leq x \leq 3$ is the solution. Test some values.

$x = -1$	$x = 0$	$x = 4$
$\sqrt{20(-1) + 4}$ is not a real number, so the inequality is not satisfied.	$5 - \sqrt{20(0) + 4} = 3$, so the inequality is satisfied.	$5 - \sqrt{20(4) + 4} \approx -4.2$, so the inequality is not satisfied

Therefore the solution $-\dfrac{1}{5} \leq x \leq 3$ checks.

Exercises

Solve each inequality.

1. $\sqrt{c - 2} + 4 \geq 7$

2. $3\sqrt{2x - 1} + 6 < 15$

3. $\sqrt{10x + 9} - 2 > 5$

4. $5\sqrt[3]{x + 2} - 8 < 2$

5. $8 - \sqrt{3x + 4} \geq 3$

6. $\sqrt{2x + 8} - 4 > 2$

7. $9 - \sqrt{6x + 3} \geq 6$

8. $\dfrac{20}{\sqrt{3x + 1}} \leq 4$

9. $2\sqrt{5x - 6} - 1 < 5$

10. $\sqrt{2x + 12} + 4 \geq 12$

11. $\sqrt{2d + 1} + \sqrt{d} \leq 5$

12. $4\sqrt{b + 3} - \sqrt{b - 2} \geq 10$

8-1 Study Guide and Intervention

Multiplying and Dividing Rational Expressions

Simplify Rational Expressions A ratio of two polynomial expressions is a **rational expression**. To simplify a rational expression, divide both the numerator and the denominator by their greatest common factor (GCF).

Multiplying Rational Expressions	For all rational expressions $\frac{a}{b}$ and $\frac{c}{d}$, $\frac{a}{b} \cdot \frac{c}{d} = \frac{ac}{bd}$, if $b \neq 0$ and $d \neq 0$.
Dividing Rational Expressions	For all rational expressions $\frac{a}{b}$ and $\frac{c}{d}$, $\frac{a}{b} \div \frac{c}{d} = \frac{ad}{bc}$, if $b \neq 0$, $c \neq 0$, and $d \neq 0$.

Example Simplify each expression.

a. $\dfrac{24a^5b^2}{(2ab)^4}$

$$\frac{24a^5b^2}{(2ab)^4} = \frac{2 \cdot 2 \cdot 2 \cdot 3 \cdot a \cdot a \cdot a \cdot a \cdot a \cdot b \cdot b}{2 \cdot 2 \cdot 2 \cdot 2 \cdot a \cdot a \cdot a \cdot a \cdot b \cdot b \cdot b \cdot b} = \frac{3a}{2b^2}$$

b. $\dfrac{3r^2s^3}{5t^4} \cdot \dfrac{20t^2}{9r^3s}$

$$\frac{3r^2s^3}{5t^4} \cdot \frac{20t^2}{9r^3s} = \frac{3 \cdot r \cdot r \cdot s \cdot s \cdot s \cdot 2 \cdot 2 \cdot 5 \cdot t \cdot t}{5 \cdot t \cdot t \cdot t \cdot t \cdot 3 \cdot 3 \cdot r \cdot r \cdot r \cdot s} = \frac{2 \cdot 2 \cdot s \cdot s}{3 \cdot r \cdot t \cdot t} = \frac{4s^2}{3rt^2}$$

c. $\dfrac{x^2 + 8x + 16}{2x - 2} \div \dfrac{x^2 + 2x - 8}{x - 1}$

$$\frac{x^2 + 8x + 16}{2x - 2} \div \frac{x^2 + 2x - 8}{x - 1} = \frac{x^2 + 8x + 16}{2x - 2} \cdot \frac{x - 1}{x^2 + 2x - 8}$$

$$= \frac{(x + 4)(x + 4)(x - 1)}{2(x - 1)(x - 2)(x + 4)} = \frac{x + 4}{2(x - 2)}$$

Exercises

Simplify each expression.

1. $\dfrac{(-2ab^2)^3}{20ab^4}$

2. $\dfrac{4x^2 - 12x + 9}{9 - 6x}$

3. $\dfrac{x^2 + x - 6}{x^2 - 6x - 27}$

4. $\dfrac{3m^3 - 3m}{6m^4} \cdot \dfrac{4m^5}{m + 1}$

5. $\dfrac{c^2 - 3c}{c^2 - 25} \cdot \dfrac{c^2 + 4c - 5}{c^2 - 4c + 3}$

6. $\dfrac{(m - 3)^2}{m^2 - 6m + 9} \cdot \dfrac{m^3 - 9m}{m^2 - 9}$

7. $\dfrac{6xy^4}{25z^3} \div \dfrac{18xz^2}{5y}$

8. $\dfrac{16p^2 - 8p + 1}{14p^4} \div \dfrac{4p^2 + 7p - 2}{7p^5}$

9. $\dfrac{2m - 1}{m^2 - 3m - 10} \div \dfrac{4m^2 - 1}{4m + 8}$

8-1 Study Guide and Intervention (continued)

Multiplying and Dividing Rational Expressions

Simplify Complex Fractions A **complex fraction** is a rational expression whose numerator and/or denominator contains a rational expression. To simplify a complex fraction, first rewrite it as a division problem.

Example Simplify $\dfrac{\dfrac{3s-1}{s}}{\dfrac{3s^2+8s-3}{s^4}}$.

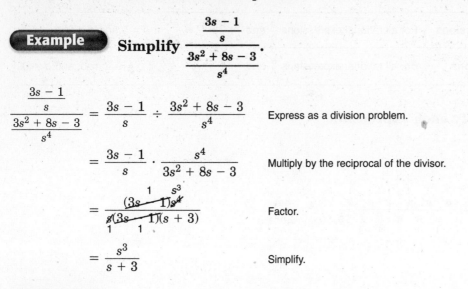

$\dfrac{\dfrac{3s-1}{s}}{\dfrac{3s^2+8s-3}{s^4}} = \dfrac{3s-1}{s} \div \dfrac{3s^2+8s-3}{s^4}$ Express as a division problem.

$= \dfrac{3s-1}{s} \cdot \dfrac{s^4}{3s^2+8s-3}$ Multiply by the reciprocal of the divisor.

$= \dfrac{(3s-1)s^4}{s(3s-1)(s+3)}$ Factor.

$= \dfrac{s^3}{s+3}$ Simplify.

Exercises

Simplify.

1. $\dfrac{\dfrac{x^3y^2z}{a^2b^2}}{\dfrac{a^3x^2y}{b^2}}$

2. $\dfrac{\dfrac{a^2bc^3}{x^2y^2}}{\dfrac{ab^2}{c^4x^2y}}$

3. $\dfrac{\dfrac{b^2-1}{3b+2}}{\dfrac{b+1}{3b^2-b-2}}$

4. $\dfrac{\dfrac{b^2-100}{b^2}}{\dfrac{3b^2-31b+10}{2b}}$

5. $\dfrac{\dfrac{x-4}{x^2+6x+9}}{\dfrac{x^2-2x-8}{3+x}}$

6. $\dfrac{\dfrac{a^2-16}{a+2}}{\dfrac{a^2+3a-4}{a^2+a-2}}$

7. $\dfrac{\dfrac{2x^2+9x+9}{x+1}}{\dfrac{10x^2+19x+6}{5x^2+7x+2}}$

8. $\dfrac{\dfrac{b+2}{b^2-6b+8}}{\dfrac{b^2+b-2}{b^2-16}}$

9. $\dfrac{\dfrac{x^2-x-2}{x^3+6x^2-x-30}}{\dfrac{x+1}{x+3}}$

8-2 Study Guide and Intervention

Adding and Subtracting Rational Expressions

LCM of Polynomials To find the least common multiple of two or more polynomials, factor each expression. The LCM contains each factor the greatest number of times it appears as a factor.

Example Find the LCM of $16p^2q^3r$, $40pq^4r^2$, and $15p^3r^4$.

$16p^2q^3r = 2^4 \cdot p^2 \cdot q^3 \cdot r$
$40pq^4r^2 = 2^3 \cdot 5 \cdot p \cdot q^4 \cdot r^2$
$15p^3r^4 = 3 \cdot 5 \cdot p^3 \cdot r^4$
$LCM = 2^4 \cdot 3 \cdot 5 \cdot p^3 \cdot q^4 \cdot r^4$
$\quad = 240p^3q^4r^4$

Example Find the LCM of $3m^2 - 3m - 6$ and $4m^2 + 12m - 40$.

$3m^2 - 3m - 6 = 3(m + 1)(m - 2)$
$4m^2 + 12m - 40 = 4(m - 2)(m + 5)$
$LCM = 12(m + 1)(m - 2)(m + 5)$

Exercises

Find the LCM of each set of polynomials.

1. $14ab^2$, $42bc^3$, $18a^2c$

2. $8cdf^3$, $28c^2f$, $35d^4f^2$

3. $65x^4y$, $10x^2y^2$, $26y^4$

4. $11mn^5$, $18m^2n^3$, $20mn^4$

5. $15a^4b$, $50a^2b^2$, $40b^8$

6. $24p^7q$, $30p^2q^2$, $45pq^3$

7. $39b^2c^2$, $52b^4c$, $12c^3$

8. $12xy^4$, $42x^2y$, $30x^2y^3$

9. $56stv^2$, $24s^2v^2$, $70t^3v^3$

10. $x^2 + 3x$, $10x^2 + 25x - 15$

11. $9x^2 - 12x + 4$, $3x^2 + 10x - 8$

12. $22x^2 + 66x - 220$, $4x^2 - 16$

13. $8x^2 - 36x - 20$, $2x^2 + 2x - 60$

14. $5x^2 - 125$, $5x^2 + 24x - 5$

15. $3x^2 - 18x + 27$, $2x^3 - 4x^2 - 6x$

16. $45x^2 - 6x - 3$, $45x^2 - 5$

17. $x^3 + 4x^2 - x - 4$, $x^2 + 2x - 3$

18. $54x^3 - 24x$, $12x^2 - 26x + 12$

8-2 Study Guide and Intervention *(continued)*

Adding and Subtracting Rational Expressions

Add and Subtract Rational Expressions To add or subtract rational expressions, follow these steps.

> **Step 1** If necessary, find equivalent fractions that have the same denominator.
> **Step 2** Add or subtract the numerators.
> **Step 3** Combine any like terms in the numerator.
> **Step 4** Factor if possible.
> **Step 5** Simplify if possible.

Example Simplify $\dfrac{6}{2x^2 + 2x - 12} - \dfrac{2}{x^2 - 4}$.

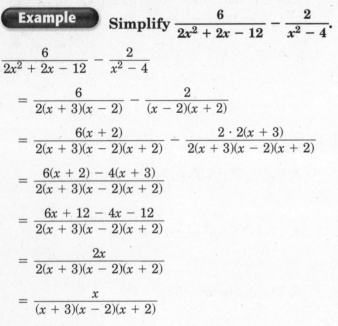

$$\frac{6}{2x^2 + 2x - 12} - \frac{2}{x^2 - 4}$$

$$= \frac{6}{2(x + 3)(x - 2)} - \frac{2}{(x - 2)(x + 2)} \qquad \text{Factor the denominators.}$$

$$= \frac{6(x + 2)}{2(x + 3)(x - 2)(x + 2)} - \frac{2 \cdot 2(x + 3)}{2(x + 3)(x - 2)(x + 2)} \qquad \text{The LCD is } 2(x + 3)(x - 2)(x + 2).$$

$$= \frac{6(x + 2) - 4(x + 3)}{2(x + 3)(x - 2)(x + 2)} \qquad \text{Subtract the numerators.}$$

$$= \frac{6x + 12 - 4x - 12}{2(x + 3)(x - 2)(x + 2)} \qquad \text{Distributive Property}$$

$$= \frac{2x}{2(x + 3)(x - 2)(x + 2)} \qquad \text{Combine like terms.}$$

$$= \frac{x}{(x + 3)(x - 2)(x + 2)} \qquad \text{Simplify.}$$

Exercises

Simplify each expression.

1. $\dfrac{-7xy}{3x} + \dfrac{4y^2}{2y}$

2. $\dfrac{2}{x - 3} - \dfrac{1}{x - 1}$

3. $\dfrac{4a}{3bc} - \dfrac{15b}{5ac}$

4. $\dfrac{3}{x + 2} + \dfrac{4x + 5}{3x + 6}$

5. $\dfrac{3x + 3}{x^2 + 2x + 1} + \dfrac{x - 1}{x^2 - 1}$

6. $\dfrac{4}{4x^2 - 4x + 1} - \dfrac{5x}{20x^2 - 5}$

8-3 Study Guide and Intervention

Graphing Rational Functions

Domain and Range

Rational Function	an equation of the form $f(x) = \frac{p(x)}{q(x)}$, where $p(x)$ and $q(x)$ are polynomial expressions and $q(x) \neq 0$
Domain	The domain of a rational function is limited to values for which the function is defined.
Vertical Asymptote	An asymptote is a line that the graph of a function approaches. If the simplified form of the related rational expression is undefined for $x = a$, then $x = a$ is a vertical asymptote.
Point Discontinuity	Point discontinuity is like a hole in a graph. If the original related expression is undefined for $x = a$ but the simplified expression is defined for $x = a$, then there is a hole in the graph at $x = a$.
Horizontal Asymptote	Often a horizontal asymptote occurs in the graph of a rational function where a value is excluded from the range.

Example Determine the equations of any vertical asymptotes and the values of x for any holes in the graph of $f(x) = \frac{4x^2 + x - 3}{x^2 - 1}$.

First factor the numerator and the denominator of the rational expression.

$f(x) = \frac{4x^2 + x - 3}{x^2 - 1} = \frac{(4x - 3)(x + 1)}{(x + 1)(x - 1)}$

The function is undefined for $x = 1$ and $x = -1$.

Since $\frac{(4x - 3)(x + 1)}{(x + 1)(x - 1)} = \frac{4x - 3}{x - 1}$, $x = 1$ is a vertical asymptote. The simplified expression is defined for $x = -1$, so this value represents a hole in the graph.

Exercises

Determine the equations of any vertical asymptotes and the values of x for any holes in the graph of each rational function.

1. $f(x) = \dfrac{4}{x^2 + 3x - 10}$

2. $f(x) = \dfrac{2x^2 - x - 10}{2x - 5}$

3. $f(x) = \dfrac{x^2 - x - 12}{x^2 - 4x}$

4. $f(x) = \dfrac{3x - 1}{3x^2 + 5x - 2}$

5. $f(x) = \dfrac{x^2 - 6x - 7}{x^2 + 6x - 7}$

6. $f(x) = \dfrac{3x^2 - 5x - 2}{x + 3}$

7. $f(x) = \dfrac{x + 1}{x^2 - 6x + 5}$

8. $f(x) = \dfrac{2x^2 - x - 3}{2x^2 + 3x - 9}$

9. $f(x) = \dfrac{x^3 - 2x^2 - 5x + 6}{x^2 - 4x + 3}$

8-3 Study Guide and Intervention (continued)

Graphing Rational Functions

Graph Rational Functions Use the following steps to graph a rational function.

Step 1	First see if the function has any vertical asymptotes or point discontinuities.
Step 2	Draw any vertical asymptotes.
Step 3	Make a table of values.
Step 4	Plot the points and draw the graph.

Example Graph $f(x) = \dfrac{x-1}{x^2+2x-3}$.

$\dfrac{x-1}{x^2+2x-3} = \dfrac{x-1}{(x-1)(x+3)}$ or $\dfrac{1}{x+3}$

Therefore the graph of $f(x)$ has an asymptote at $x = -3$ and a point discontinuity at $x = 1$.

Make a table of values. Plot the points and draw the graph.

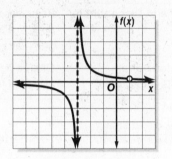

x	-2.5	-2	-1	-3.5	-4	-5
f(x)	2	1	0.5	-2	-1	-0.5

Exercises

Graph each rational function.

1. $f(x) = \dfrac{3}{x+1}$

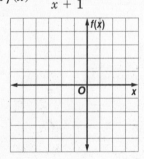

2. $f(x) = \dfrac{2}{x}$

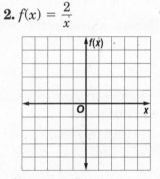

3. $f(x) = \dfrac{2x+1}{x-3}$

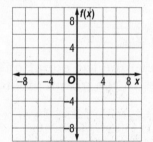

4. $f(x) = \dfrac{2}{(x+3)^2}$

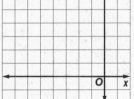

5. $f(x) = \dfrac{x^2-x-6}{x-3}$

6. $f(x) = \dfrac{x^2-6x+8}{x^2-x-2}$

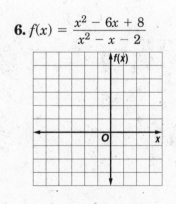

8-4 Study Guide and Intervention

Direct, Joint, and Inverse Variation

Direct Variation and Joint Variation

Direct Variation	y varies directly as x if there is some nonzero constant k such that $y = kx$. k is called the constant of variation.
Joint Variation	y varies jointly as x and z if there is some number k such that $y = kxz$, where $x \neq 0$ and $z \neq 0$.

Example Find each value.

a. If y varies directly as x and $y = 16$ when $x = 4$, find x when $y = 20$.

$$\frac{y_1}{x_1} = \frac{y_2}{x_2} \quad \text{Direct proportion}$$

$$\frac{16}{4} = \frac{20}{x_2} \quad y_1 = 16,\ x_1 = 4,\ \text{and } y_2 = 20$$

$$16x_2 = (20)(4) \quad \text{Cross multiply.}$$

$$x_2 = 5 \quad \text{Simplify.}$$

The value of x is 5 when y is 20.

b. If y varies jointly as x and z and $y = 10$ when $x = 2$ and $z = 4$, find y when $x = 4$ and $z = 3$.

$$\frac{y_1}{x_1 z_1} = \frac{y_2}{x_2 z_2} \quad \text{Joint variation}$$

$$\frac{10}{2 \cdot 4} = \frac{y_2}{4 \cdot 3} \quad y_1 = 10,\ x_1 = 2,\ z_1 = 4,\ x_2 = 4,\ \text{and } z_2 = 3$$

$$120 = 8y_2 \quad \text{Simplify.}$$

$$y_2 = 15 \quad \text{Divide each side by 8.}$$

The value of y is 15 when $x = 4$ and $z = 3$.

Exercises

Find each value.

1. If y varies directly as x and $y = 9$ when $x = 6$, find y when $x = 8$.

2. If y varies directly as x and $y = 16$ when $x = 36$, find y when $x = 54$.

3. If y varies directly as x and $x = 15$ when $y = 5$, find x when $y = 9$.

4. If y varies directly as x and $x = 33$ when $y = 22$, find x when $y = 32$.

5. Suppose y varies jointly as x and z. Find y when $x = 5$ and $z = 3$, if $y = 18$ when $x = 3$ and $z = 2$.

6. Suppose y varies jointly as x and z. Find y when $x = 6$ and $z = 8$, if $y = 6$ when $x = 4$ and $z = 2$.

7. Suppose y varies jointly as x and z. Find y when $x = 4$ and $z = 11$, if $y = 60$ when $x = 3$ and $z = 5$.

8. Suppose y varies jointly as x and z. Find y when $x = 5$ and $z = 2$, if $y = 84$ when $x = 4$ and $z = 7$.

9. If y varies directly as x and $y = 39$ when $x = 52$, find y when $x = 22$.

10. If y varies directly as x and $x = 60$ when $y = 75$; find x when $y = 42$.

11. Suppose y varies jointly as x and z. Find y when $x = 7$ and $z = 18$, if $y = 351$ when $x = 6$ and $z = 13$.

12. Suppose y varies jointly as x and z. Find y when $x = 5$ and $z = 27$, if $y = 480$ when $x = 9$ and $z = 20$.

8-4 Study Guide and Intervention (continued)

Direct, Joint, and Inverse Variation

Inverse Variation

Inverse Variation	y varies inversely as x if there is some nonzero constant k such that $xy = k$ or $y = \dfrac{k}{x}$.

Example If a varies inversely as b and $a = 8$ when $b = 12$, find a when $b = 4$.

$\dfrac{a_1}{b_2} = \dfrac{a_2}{b_1}$ Inverse variation

$\dfrac{8}{4} = \dfrac{a_2}{12}$ $a_1 = 8,\ b_1 = 12,\ b_2 = 4$

$8(12) = 4a_2$ Cross multiply.

$96 = 4a_2$ Simplify.

$24 = a_2$ Divide each side by 4.

When $b = 4$, the value of a is 24.

Exercises

Find each value.

1. If y varies inversely as x and $y = 12$ when $x = 10$, find y when $x = 15$.

2. If y varies inversely as x and $y = 100$ when $x = 38$, find y when $x = 76$.

3. If y varies inversely as x and $y = 32$ when $x = 42$, find y when $x = 24$.

4. If y varies inversely as x and $y = 36$ when $x = 10$, find y when $x = 30$.

5. If y varies inversely as x and $y = 18$ when $x = 124$, find y when $x = 93$.

6. If y varies inversely as x and $y = 90$ when $x = 35$, find y when $x = 50$.

7. If y varies inversely as x and $y = 42$ when $x = 48$, find y when $x = 36$.

8. If y varies inversely as x and $y = 44$ when $x = 20$, find y when $x = 55$.

9. If y varies inversely as x and $y = 80$ when $x = 14$, find y when $x = 35$.

10. If y varies inversely as x and $y = 3$ when $x = 8$, find y when $x = 40$.

11. If y varies inversely as x and $y = 16$ when $x = 42$, find y when $x = 14$.

12. If y varies inversely as x and $y = 23$ when $x = 12$, find y when $x = 15$.

8-5 Study Guide and Intervention

Classes of Functions

Identify Graphs You should be familiar with the graphs of the following functions.

Function	Description of Graph
Constant	a horizontal line that crosses the y-axis at a
Direct Variation	a line that passes through the origin and is neither horizontal nor vertical
Identity	a line that passes through the point (a, a), where a is any real number
Greatest Integer	a step function
Absolute Value	V-shaped graph
Quadratic	a parabola
Square Root	a curve that starts at a point and curves in only one direction
Rational	a graph with one or more asymptotes and/or holes
Inverse Variation	a graph with 2 curved branches and 2 asymptotes, x = 0 and y = 0 (special case of rational function)

Exercises

Identify the function represented by each graph.

1.
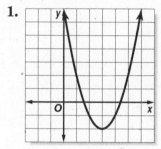

2.
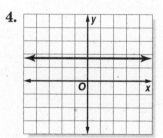

3.
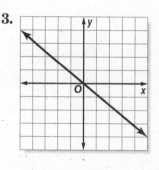

4.
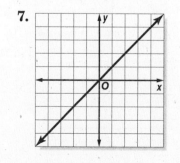

5.
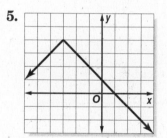

6.
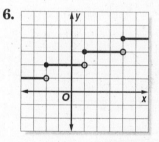

7.
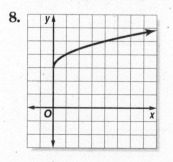

8.
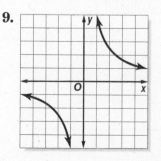

9.

8-5 Study Guide and Intervention (continued)

Classes of Functions

Identify Equations You should be able to graph the equations of the following functions.

Function	General Equation
Constant	$y = a$
Direct Variation	$y = ax$
Greatest Integer	equation includes a variable within the greatest integer symbol, $[\![\ ]\!]$
Absolute Value	equation includes a variable within the absolute value symbol, $|\ |$
Quadratic	$y = ax^2 + bx + c$, where $a \neq 0$
Square Root	equation includes a variable beneath the radical sign, $\sqrt{\ }$
Rational	$y = \dfrac{p(x)}{q(x)}$
Inverse Variation	$y = \dfrac{a}{x}$

Exercises

Identify the function represented by each equation. Then graph the equation.

1. $y = \dfrac{6}{x}$

2. $y = \dfrac{4}{3}x$

3. $y = -\dfrac{x^2}{2}$

4. $y = [\![3x]\!] - 1$

5. $y = -\dfrac{2}{x}$

6. $y = \left[\!\left[\dfrac{x}{2}\right]\!\right]$

7. $y = \sqrt{x - 2}$

8. $y = 3.2$

9. $y = \dfrac{x^2 + 5x + 6}{x + 2}$

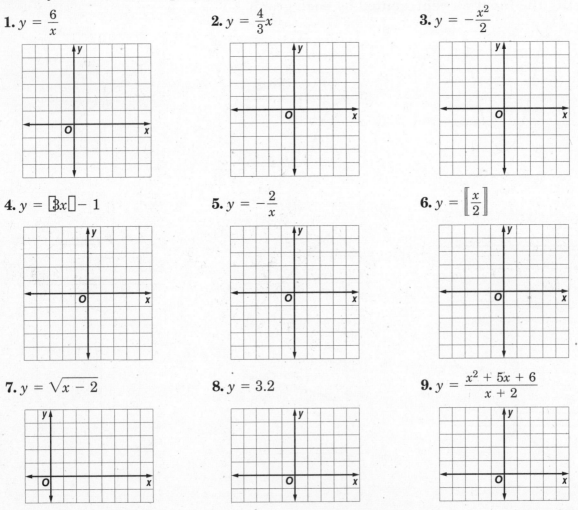

8-6 Study Guide and Intervention

Solving Rational Equations and Inequalities

Solve Rational Equations A **rational equation** contains one or more rational expressions. To solve a rational equation, first multiply each side by the least common denominator of all of the denominators. Be sure to exclude any solution that would produce a denominator of zero.

Example Solve $\dfrac{9}{10} + \dfrac{2}{x+1} = \dfrac{2}{5}$.

$$\dfrac{9}{10} + \dfrac{2}{x+1} = \dfrac{2}{5} \qquad \text{Original equation}$$

$$10(x+1)\left(\dfrac{9}{10} + \dfrac{2}{x+1}\right) = 10(x+1)\left(\dfrac{2}{5}\right) \qquad \text{Multiply each side by } 10(x+1).$$

$$9(x+1) + 2(10) = 4(x+1) \qquad \text{Multiply.}$$

$$9x + 9 + 20 = 4x + 4 \qquad \text{Distributive Property}$$

$$5x = -25 \qquad \text{Subtract 4x and 29 from each side.}$$

$$x = -5 \qquad \text{Divide each side by 5.}$$

Check $\qquad \dfrac{9}{10} + \dfrac{2}{x+1} = \dfrac{2}{5}$ Original equation

$$\dfrac{9}{10} + \dfrac{2}{-5+1} \stackrel{?}{=} \dfrac{2}{5} \quad x = -5$$

$$\dfrac{18}{20} - \dfrac{10}{20} \stackrel{?}{=} \dfrac{2}{5} \quad \text{Simplify.}$$

$$\dfrac{2}{5} = \dfrac{2}{5}$$

Exercises

Solve each equation.

1. $\dfrac{2y}{3} - \dfrac{y+3}{6} = 2$

2. $\dfrac{4t-3}{5} - \dfrac{4-2t}{3} = 1$

3. $\dfrac{2x+1}{3} - \dfrac{x-5}{4} = \dfrac{1}{2}$

4. $\dfrac{3m+2}{5m} + \dfrac{2m-1}{2m} = 4$

5. $\dfrac{4}{x-1} = \dfrac{x+1}{12}$

6. $\dfrac{x}{x-2} + \dfrac{4}{x-2} = 10$

7. **NAVIGATION** The current in a river is 6 miles per hour. In her motorboat Marissa can travel 12 miles upstream or 16 miles downstream in the same amount of time. What is the speed of her motorboat in still water? Is this a reasonable answer? Explain.

8. **WORK** Adam, Bethany, and Carlos own a painting company. To paint a particular house alone, Adam estimates that it would take him 4 days, Bethany estimates $5\dfrac{1}{2}$ days, and Carlos 6 days. If these estimates are accurate, how long should it take the three of them to paint the house if they work together? Is this a reasonable answer?

8-6 Study Guide and Intervention (continued)

Solving Rational Equations and Inequalities

Solve Rational Inequalities To solve a rational inequality, complete the following steps.

Step 1	State the excluded values.
Step 2	Solve the related equation.
Step 3	Use the values from steps 1 and 2 to divide the number line into regions. Test a value in each region to see which regions satisfy the original inequality.

Example Solve $\dfrac{2}{3n} + \dfrac{4}{5n} \le \dfrac{2}{3}$.

Step 1 The value of 0 is excluded since this value would result in a denominator of 0.

Step 2 Solve the related equation.

$$\dfrac{2}{3n} + \dfrac{4}{5n} = \dfrac{2}{3} \qquad \text{Related equation}$$

$$15n\left(\dfrac{2}{3n} + \dfrac{4}{5n}\right) = 15n\left(\dfrac{2}{3}\right) \qquad \text{Multiply each side by } 15n.$$

$$10 + 12 = 10n \qquad \text{Simplify.}$$

$$22 = 10n \qquad \text{Simplify.}$$

$$2.2 = n \qquad \text{Simplify.}$$

Step 3 Draw a number with vertical lines at the excluded value and the solution to the equation.

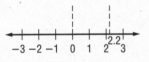

Test $n = -1$. $\qquad$ Test $n = 1$. $\qquad$ Test $n = 3$.

$-\dfrac{2}{3} + \left(-\dfrac{4}{5}\right) \le \dfrac{2}{3}$ is true. $\quad$ $\dfrac{2}{3} + \dfrac{4}{5} \le \dfrac{2}{3}$ is *not* true. $\quad$ $\dfrac{2}{9} + \dfrac{4}{15} \le \dfrac{2}{3}$ is true.

The solution is $n < 0$ or $n \ge 2.2$.

Exercises

Solve each inequality.

1. $\dfrac{3}{a+1} \ge 3$

2. $\dfrac{1}{x} \ge 4x$

3. $\dfrac{1}{2p} + \dfrac{4}{5p} > \dfrac{2}{3}$

4. $\dfrac{3}{2x} - \dfrac{2}{x} > \dfrac{1}{4}$

5. $\dfrac{4}{x-1} + \dfrac{5}{x} < 2$

6. $\dfrac{3}{x^2-1} + 1 > \dfrac{2}{x-1}$

NAME _____ DATE _____ PERIOD _____

9-1 Study Guide and Intervention

Exponential Functions

Exponential Functions An **exponential function** has the form $y = ab^x$, where $a \neq 0$, $b > 0$, and $b \neq 1$.

Properties of an Exponential Function	1. The function is continuous and one-to-one. 2. The domain is the set of all real numbers. 3. The x-axis is the asymptote of the graph. 4. The range is the set of all positive numbers if $a > 0$ and all negative numbers if $a < 0$. 5. The graph contains the point $(0, a)$.
Exponential Growth and Decay	If $a > 0$ and $b > 1$, the function $y = ab^x$ represents exponential growth. If $a > 0$ and $0 < b < 1$, the function $y = ab^x$ represents exponential decay.

Example 1 Sketch the graph of $y = 0.1(4)^x$. Then state the function's domain and range.

Make a table of values. Connect the points to form a smooth curve.

x	−1	0	1	2	3
y	0.025	0.1	0.4	1.6	6.4

The domain of the function is all real numbers, while the range is the set of all positive real numbers.

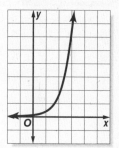

Example 2 Determine whether each function represents exponential *growth, decay,* or *neither.*

a. $y = 0.5(2)^x$

exponential growth, since the base, 2, is greater than 1

b. $y = -2.8(2)^x$

neither, since −2.8, the value of a is less than 0.

c. $y = 1.1(0.5)^x$

exponential decay, since the base, 0.5, is between 0 and 1

Exercises

Sketch the graph of each function. Then state the function's domain and range.

1. $y = 3(2)^x$

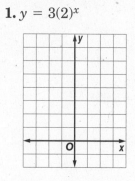

2. $y = -2\left(\dfrac{1}{4}\right)^x$

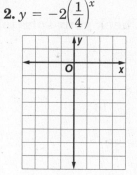

3. $y = 0.25(5)^x$

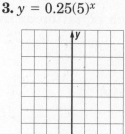

Determine whether each function represents exponential growth, decay, or neither.

4. $y = 0.3(1.2)^x$

5. $y = -5\left(\dfrac{4}{5}\right)^x$

6. $y = 3(10)^{-x}$

9-1 Study Guide and Intervention *(continued)*

Exponential Functions

Exponential Equations and Inequalities All the properties of rational exponents that you know also apply to real exponents. Remember that $a^m \cdot a^n = a^{m+n}$, $(a^m)^n = a^{mn}$, and $a^m \div a^n = a^{m-n}$.

Property of Equality for Exponential Functions	If b is a positive number other than 1, then $b^x = b^y$ if and only if $x = y$.
Property of Inequality for Exponential Functions	If $b > 1$ then $b^x > b^y$ if and only if $x > y$ and $b^x < b^y$ if and only if $x < y$.

Example 1 Solve $4^{x-1} = 2^{x+5}$.

$4^{x-1} = 2^{x+5}$ Original equation
$(2^2)^{x-1} = 2^{x+5}$ Rewrite 4 as 2^2.
$2(x-1) = x + 5$ Prop. of Inequality for Exponential Functions
$2x - 2 = x + 5$ Distributive Property
$x = 7$ Subtract x and add 2 to each side.

Example 2 Solve $5^{2x-1} > \dfrac{1}{125}$.

$5^{2x-1} > \dfrac{1}{125}$ Original inequality
$5^{2x-1} > 5^{-3}$ Rewrite $\frac{1}{125}$ as 5^{-3}.
$2x - 1 > -3$ Prop. of Inequality for Exponential Functions
$2x > -2$ Add 1 to each side.
$x > -1$ Divide each side by 2.
The solution set is $\{x \mid x > -1\}$.

Exercises

Simplify each expression.

1. $\left(3^{\sqrt{2}}\right)^{\sqrt{2}}$

2. $25^{\sqrt{2}} \cdot 125^{\sqrt{2}}$

3. $\left(x^{\sqrt{2}}y^{3\sqrt{2}}\right)^{\sqrt{2}}$

4. $\left(x^{\sqrt{6}}\right)\left(x^{\sqrt{5}}\right)$

5. $\left(x^{\sqrt{6}}\right)^{\sqrt{5}}$

6. $(2x^{\pi})(5x^{3\pi})$

Solve each equation or inequality. Check your solution.

7. $3^{2x-1} = 3^{x+2}$

8. $2^{3x} = 4^{x+2}$

9. $3^{2x-1} = \dfrac{1}{9}$

10. $4^{x+1} = 8^{2x+3}$

11. $8^{x-2} = \dfrac{1}{16}$

12. $25^{2x} = 125^{x+2}$

13. $4^{\sqrt{x}} = 16^{\sqrt{5}}$

14. $x^{\sqrt{3}} = 36^{\sqrt{\frac{3}{4}}}$

15. $x^{\sqrt{2}} = 81^{\frac{1}{\sqrt{8}}}$

16. $3^{x-4} < \dfrac{1}{27}$

17. $4^{2x-2} > 2^{x+1}$

18. $5^{2x} < 125^{x-5}$

19. $10^{4x+1} > 100^{x-2}$

20. $7^{3x} < 49^{x^2}$

21. $8^{2x-5} < 4^{x+8}$

9-2 Study Guide and Intervention

Logarithms and Logarithmic Functions

Logarithmic Functions and Expressions

Definition of Logarithm with Base b	Let b and x be positive numbers, $b \neq 1$. The logarithm of x with base b is denoted $\log_b x$ and is defined as the exponent y that makes the equation $b^y = x$ true.

The inverse of the exponential function $y = b^x$ is the **logarithmic function** $x = b^y$. This function is usually written as $y = \log_b x$.

Properties of Logarithmic Functions	1. The function is continuous and one-to-one.
	2. The domain is the set of all positive real numbers.
	3. The y-axis is an asymptote of the graph.
	4. The range is the set of all real numbers.
	5. The graph contains the point (1, 0).

Example 1 Write an exponential equation equivalent to $\log_3 243 = 5$.

$3^5 = 243$

Example 2 Write a logarithmic equation equivalent to $6^{-3} = \dfrac{1}{216}$.

$\log_6 \dfrac{1}{216} = -3$

Example 3 Evaluate $\log_8 16$.

$8^{\frac{4}{3}} = 16$, so $\log_8 16 = \dfrac{4}{3}$.

Exercises

Write each equation in logarithmic form.

1. $2^7 = 128$

2. $3^{-4} = \dfrac{1}{81}$

3. $\left(\dfrac{1}{7}\right)^3 = \dfrac{1}{343}$

Write each equation in exponential form.

4. $\log_{15} 225 = 2$

5. $\log_3 \dfrac{1}{27} = -3$

6. $\log_4 32 = \dfrac{5}{2}$

Evaluate each expression.

7. $\log_4 64$

8. $\log_2 64$

9. $\log_{100} 100{,}000$

10. $\log_5 625$

11. $\log_{27} 81$

12. $\log_{25} 5$

13. $\log_2 \dfrac{1}{128}$

14. $\log_{10} 0.00001$

15. $\log_4 \dfrac{1}{32}$

9-2 Study Guide and Intervention *(continued)*

Logarithms and Logarithmic Functions

Solve Logarithmic Equations and Inequalities

Logarithmic to Exponential Inequality	If $b > 1$, $x > 0$, and $\log_b x > y$, then $x > b^y$. If $b > 1$, $x > 0$, and $\log_b x < y$, then $0 < x < b^y$.
Property of Equality for Logarithmic Functions	If b is a positive number other than 1, then $\log_b x = \log_b y$ if and only if $x = y$.
Property of Inequality for Logarithmic Functions	If $b > 1$, then $\log_b x > \log_b y$ if and only if $x > y$, and $\log_b x < \log_b y$ if and only if $x < y$.

Example 1 Solve $\log_2 2x = 3$.

$\log_2 2x = 3$ Original equation
$2x = 2^3$ Definition of logarithm
$2x = 8$ Simplify.
$x = 4$ Simplify.
The solution is $x = 4$.

Example 2 Solve $\log_5 (4x - 3) < 3$.

$\log_5 (4x - 3) < 3$ Original equation
$0 < 4x - 3 < 5^3$ Logarithmic to exponential inequality
$3 < 4x < 125 + 3$ Addition Property of Inequalities
$\dfrac{3}{4} < x < 32$ Simplify.

The solution set is $\left\{ x \Big| \dfrac{3}{4} < x < 32 \right\}$.

Exercises

Solve each equation or inequality.

1. $\log_2 32 = 3x$

2. $\log_3 2c = -2$

3. $\log_{2x} 16 = -2$

4. $\log_{25} \left(\dfrac{x}{2} \right) = \dfrac{1}{2}$

5. $\log_4 (5x + 1) = 2$

6. $\log_8 (x - 5) = \dfrac{2}{3}$

7. $\log_4 (3x - 1) = \log_4 (2x + 3)$

8. $\log_2 (x^2 - 6) = \log_2 (2x + 2)$

9. $\log_{x + 4} 27 = 3$

10. $\log_2 (x + 3) = 4$

11. $\log_x 1000 = 3$

12. $\log_8 (4x + 4) = 2$

13. $\log_2 2x > 2$

14. $\log_5 x > 2$

15. $\log_2 (3x + 1) < 4$

16. $\log_4 (2x) > -\dfrac{1}{2}$

17. $\log_3 (x + 3) < 3$

18. $\log_{27} 6x > \dfrac{2}{3}$

9-3 Study Guide and Intervention

Properties of Logarithms

Properties of Logarithms Properties of exponents can be used to develop the following properties of logarithms.

Product Property of Logarithms	For all positive numbers m, n, and b, where $b \neq 1$, $\log_b mn = \log_b m + \log_b n$.
Quotient Property of Logarithms	For all positive numbers m, n, and b, where $b \neq 1$, $\log_b \frac{m}{n} = \log_b m - \log_b n$.
Power Property of Logarithms	For any real number p and positive numbers m and b, where $b \neq 1$, $\log_b m^p = p \log_b m$.

Example Use $\log_3 28 \approx 3.0331$ and $\log_3 4 \approx 1.2619$ to approximate the value of each expression.

a. $\log_3 36$

$$\log_3 36 = \log_3 (3^2 \cdot 4)$$
$$= \log_3 3^2 + \log_3 4$$
$$= 2 + \log_3 4$$
$$\approx 2 + 1.2619$$
$$\approx 3.2619$$

b. $\log_3 7$

$$\log_3 7 = \log_3 \left(\frac{28}{4}\right)$$
$$= \log_3 28 - \log_3 4$$
$$\approx 3.0331 - 1.2619$$
$$\approx 1.7712$$

c. $\log_3 256$

$$\log_3 256 = \log_3 (4^4)$$
$$= 4 \cdot \log_3 4$$
$$\approx 4(1.2619)$$
$$\approx 5.0476$$

Exercises

Use $\log_{12} 3 \approx 0.4421$ and $\log_{12} 7 \approx 0.7831$ to evaluate each expression.

1. $\log_{12} 21$

2. $\log_{12} \frac{7}{3}$

3. $\log_{12} 49$

4. $\log_{12} 36$

5. $\log_{12} 63$

6. $\log_{12} \frac{27}{49}$

7. $\log_{12} \frac{81}{49}$

8. $\log_{12} 16{,}807$

9. $\log_{12} 441$

Use $\log_5 3 \approx 0.6826$ and $\log_5 4 \approx 0.8614$ to evaluate each expression.

10. $\log_5 12$

11. $\log_5 100$

12. $\log_5 0.75$

13. $\log_5 144$

14. $\log_5 \frac{27}{16}$

15. $\log_5 375$

16. $\log_5 1.\overline{3}$

17. $\log_5 \frac{9}{16}$

18. $\log_5 \frac{81}{5}$

9-3 Study Guide and Intervention (continued)

Properties of Logarithms

Solve Logarithmic Equations You can use the properties of logarithms to solve equations involving logarithms.

Example Solve each equation.

a. $2 \log_3 x - \log_3 4 = \log_3 25$

$2 \log_3 x - \log_3 4 = \log_3 25$	Original equation
$\log_3 x^2 - \log_3 4 = \log_3 25$	Power Property
$\log_3 \dfrac{x^2}{4} = \log_3 25$	Quotient Property
$\dfrac{x^2}{4} = 25$	Property of Equality for Logarithmic Functions
$x^2 = 100$	Multiply each side by 4.
$x = \pm 10$	Take the square root of each side.

Since logarithms are undefined for $x < 0$, -10 is an extraneous solution. The only solution is 10.

b. $\log_2 x + \log_2 (x + 2) = 3$

$\log_2 x + \log_2 (x + 2) = 3$	Original equation
$\log_2 x(x + 2) = 3$	Product Property
$x(x + 2) = 2^3$	Definition of logarithm
$x^2 + 2x = 8$	Distributive Property
$x^2 + 2x - 8 = 0$	Subtract 8 from each side.
$(x + 4)(x - 2) = 0$	Factor.
$x = 2$ or $x = -4$	Zero Product Property

Since logarithms are undefined for $x < 0$, -4 is an extraneous solution. The only solution is 2.

Exercises

Solve each equation. Check your solutions.

1. $\log_5 4 + \log_5 2x = \log_5 24$

2. $3 \log_4 6 - \log_4 8 = \log_4 x$

3. $\dfrac{1}{2} \log_6 25 + \log_6 x = \log_6 20$

4. $\log_2 4 - \log_2 (x + 3) = \log_2 8$

5. $\log_6 2x - \log_6 3 = \log_6 (x - 1)$

6. $2 \log_4 (x + 1) = \log_4 (11 - x)$

7. $\log_2 x - 3 \log_2 5 = 2 \log_2 10$

8. $3 \log_2 x - 2 \log_2 5x = 2$

9. $\log_3 (c + 3) - \log_3 (4c - 1) = \log_3 5$

10. $\log_5 (x + 3) - \log_5 (2x - 1) = 2$

9-4 Study Guide and Intervention

Common Logarithms

Common Logarithms Base 10 logarithms are called **common logarithms**. The expression $\log_{10} x$ is usually written without the subscript as $\log x$. Use the LOG key on your calculator to evaluate common logarithms.

The relation between exponents and logarithms gives the following identity.

Inverse Property of Logarithms and Exponents	$10^{\log x} = x$

Example 1 **Evaluate log 50 to four decimal places.**

Use the LOG key on your calculator. To four decimal places, $\log 50 = 1.6990$.

Example 2 **Solve $3^{2x + 1} = 12$.**

$$3^{2x + 1} = 12 \qquad \text{Original equation}$$
$$\log 3^{2x + 1} = \log 12 \qquad \text{Property of Equality for Logarithms}$$
$$(2x + 1) \log 3 = \log 12 \qquad \text{Power Property of Logarithms}$$
$$2x + 1 = \frac{\log 12}{\log 3} \qquad \text{Divide each side by log 3.}$$
$$2x = \frac{\log 12}{\log 3} - 1 \qquad \text{Subtract 1 from each side.}$$
$$x = \frac{1}{2}\left(\frac{\log 12}{\log 3} - 1\right) \qquad \text{Multiply each side by } \frac{1}{2}.$$
$$x \approx 0.6309$$

Exercises

Use a calculator to evaluate each expression to four decimal places.

1. $\log 18$

2. $\log 39$

3. $\log 120$

4. $\log 5.8$

5. $\log 42.3$

6. $\log 0.003$

Solve each equation or inequality. Round to four decimal places.

7. $4^{3x} = 12$

8. $6^{x + 2} = 18$

9. $5^{4x - 2} = 120$

10. $7^{3x - 1} \geq 21$

11. $2.4^{x + 4} = 30$

12. $6.5^{2x} \geq 200$

13. $3.6^{4x - 1} = 85.4$

14. $2^{x + 5} = 3^{x - 2}$

15. $9^{3x} = 4^{5x + 2}$

16. $6^{x - 5} = 27^{x + 3}$

9-4 Study Guide and Intervention (continued)
Common Logarithms

Change of Base Formula The following formula is used to change expressions with different logarithmic bases to common logarithm expressions.

Change of Base Formula	For all positive numbers a, b, and n, where $a \neq 1$ and $b \neq 1$, $\log_a n = \dfrac{\log_b n}{\log_b a}$

Example Express $\log_8 15$ in terms of common logarithms. Then approximate its value to four decimal places.

$\log_8 15 = \dfrac{\log_{10} 15}{\log_{10} 8}$ Change of Base Formula

$ \approx 1.3023$ Simplify.

The value of $\log_8 15$ is approximately 1.3023.

Exercises

Express each logarithm in terms of common logarithms. Then approximate its value to four decimal places.

1. $\log_3 16$ **2.** $\log_2 40$ **3.** $\log_5 35$

4. $\log_4 22$ **5.** $\log_{12} 200$ **6.** $\log_2 50$

7. $\log_5 0.4$ **8.** $\log_3 2$ **9.** $\log_4 28.5$

10. $\log_3 (20)^2$ **11.** $\log_6 (5)^4$ **12.** $\log_8 (4)^5$

13. $\log_5 (8)^3$ **14.** $\log_2 (3.6)^6$ **15.** $\log_{12} (10.5)^4$

16. $\log_3 \sqrt{150}$ **17.** $\log_4 \sqrt[3]{39}$ **18.** $\log_5 \sqrt[4]{1600}$

9-5 Study Guide and Intervention
Base e and Natural Logarithms

Base e and Natural Logarithms The irrational number $e \approx 2.71828...$ often occurs as the base for exponential and logarithmic functions that describe real-world phenomena.

Natural Base e	As n increases, $\left(1 + \dfrac{1}{n}\right)^n$ approaches $e \approx 2.71828....$
	$\ln x = \log_e x$

The functions $y = e^x$ and $y = \ln x$ are inverse functions.

Inverse Property of Base e and Natural Logarithms	$e^{\ln x} = x$ $\ln e^x = x$

Natural base expressions can be evaluated using the e^x and ln keys on your calculator.

Example 1 Evaluate ln 1685.
Use a calculator.
$\ln 1685 \approx 7.4295$

Example 2 Write a logarithmic equation equivalent to $e^{2x} = 7$.
$e^{2x} = 7 \rightarrow \log_e 7 = 2x$ or $2x = \ln 7$

Example 3 Evaluate $\ln e^{18}$.
Use the Inverse Property of Base e and Natural Logarithms.
$\ln e^{18} = 18$

Exercises

Use a calculator to evaluate each expression to four decimal places.

1. ln 732   **2.** ln 84,350   **3.** ln 0.735   **4.** ln 100

5. ln 0.0824   **6.** ln 2.388   **7.** ln 128,245   **8.** ln 0.00614

Write an equivalent exponential or logarithmic equation.

9. $e^{15} = x$ **10.** $e^{3x} = 45$ **11.** $\ln 20 = x$ **12.** $\ln x = 8$

13. $e^{-5x} = 0.2$ **14.** $\ln (4x) = 9.6$ **15.** $e^{8.2} = 10x$ **16.** $\ln 0.0002 = x$

Evaluate each expression.

17. $\ln e^3$ **18.** $e^{\ln 42}$ **19.** $e^{\ln 0.5}$ **20.** $\ln e^{16.2}$

9-5 Study Guide and Intervention (continued)

Base e and Natural Logarithms

Equations and Inequalities with e and ln All properties of logarithms from earlier lessons can be used to solve equations and inequalities with natural logarithms.

Example Solve each equation or inequality.

a. $3e^{2x} + 2 = 10$

$3e^{2x} + 2 = 10$	Original equation
$3e^{2x} = 8$	Subtract 2 from each side.
$e^{2x} = \dfrac{8}{3}$	Divide each side by 3.
$\ln e^{2x} = \ln \dfrac{8}{3}$	Property of Equality for Logarithms
$2x = \ln \dfrac{8}{3}$	Inverse Property of Exponents and Logarithms
$x = \dfrac{1}{2} \ln \dfrac{8}{3}$	Multiply each side by $\dfrac{1}{2}$.
$x \approx 0.4904$	Use a calculator.

b. $\ln (4x - 1) < 2$

$\ln (4x - 1) < 2$	Original inequality
$e^{\ln (4x - 1)} < e^2$	Write each side using exponents and base e.
$0 < 4x - 1 < e^2$	Inverse Property of Exponents and Logarithms
$1 < 4x < e^2 + 1$	Addition Property of Inequalities
$\dfrac{1}{4} < x < \dfrac{1}{4}(e^2 + 1)$	Multiplication Property of Inequalities
$0.25 < x < 2.0973$	Use a calculator.

Exercises

Solve each equation or inequality.

1. $e^{4x} = 120$

2. $e^x \le 25$

3. $e^{x-2} + 4 = 21$

4. $\ln 6x \ge 4$

5. $\ln (x + 3) - 5 = -2$

6. $e^{-8x} \le 50$

7. $e^{4x-1} - 3 = 12$

8. $\ln (5x + 3) = 3.6$

9. $2e^{3x} + 5 = 2$

10. $6 + 3e^{x+1} = 21$

11. $\ln (2x - 5) = 8$

12. $\ln 5x + \ln 3x > 9$

9-6 Study Guide and Intervention

Exponential Growth and Decay

Exponential Decay Depreciation of value and radioactive decay are examples of **exponential decay**. When a quantity decreases by a fixed percent each time period, the amount of the quantity after t time periods is given by $y = a(1 - r)^t$, where a is the initial amount and r is the percent decrease expressed as a decimal.

Another exponential decay model often used by scientists is $y = ae^{-kt}$, where k is a constant.

Example **CONSUMER PRICES** As technology advances, the price of many technological devices such as scientific calculators and camcorders goes down. One brand of hand-held organizer sells for $89.

a. If its price decreases by 6% per year, how much will it cost after 5 years?

Use the exponential decay model with initial amount $89, percent decrease 0.06, and time 5 years.

$y = a(1 - r)^t$ Exponential decay formula

$y = 89(1 - 0.06)^5$ $a = 89, r = 0.06, t = 5$

$y = \$65.32$

After 5 years the price will be $65.32.

b. After how many years will its price be $50?

To find when the price will be $50, again use the exponential decay formula and solve for t.

$y = a(1 - r)^t$ Exponential decay formula

$50 = 89(1 - 0.06)^t$ $y = 50, a = 89, r = 0.06$

$\dfrac{50}{89} = (0.94)^t$ Divide each side by 89.

$\log\left(\dfrac{50}{89}\right) = \log(0.94)^t$ Property of Equality for Logarithms

$\log\left(\dfrac{50}{89}\right) = t \log 0.94$ Power Property

$t = \dfrac{\log\left(\dfrac{50}{89}\right)}{\log 0.94}$ Divide each side by log 0.94.

$t \approx 9.3$

The price will be $50 after about 9.3 years.

Exercises

1. **BUSINESS** A furniture store is closing out its business. Each week the owner lowers prices by 25%. After how many weeks will the sale price of a $500 item drop below $100?

CARBON DATING Use the formula $y = ae^{-0.00012t}$, where a is the initial amount of Carbon-14, t is the number of years ago the animal lived, and y is the remaining amount after t years.

2. How old is a fossil remain that has lost 95% of its Carbon-14?

3. How old is a skeleton that has 95% of its Carbon-14 remaining?

NAME _____ DATE _____ PERIOD _____

9-6 Study Guide and Intervention (continued)

Exponential Growth and Decay

Exponential Growth Population increase and growth of bacteria colonies are examples of **exponential growth**. When a quantity increases by a fixed percent each time period, the amount of that quantity after t time periods is given by $y = a(1 + r)^t$, where a is the initial amount and r is the percent increase (or rate of growth) expressed as a decimal.

Another exponential growth model often used by scientists is $y = ae^{kt}$, where k is a constant.

Example A computer engineer is hired for a salary of \$28,000. If she gets a 5% raise each year, after how many years will she be making \$50,000 or more?

Use the exponential growth model with $a = 28{,}000$, $y = 50{,}000$, and $r = 0.05$ and solve for t.

$$y = a(1 + r)^t \qquad \text{Exponential growth formula}$$

$$50{,}000 = 28{,}000(1 + 0.05)^t \qquad y = 50{,}000,\ a = 28{,}000,\ r = 0.05$$

$$\frac{50}{28} = (1.05)^t \qquad \text{Divide each side by 28,000.}$$

$$\log\left(\frac{50}{28}\right) = \log(1.05)^t \qquad \text{Property of Equality of Logarithms}$$

$$\log\left(\frac{50}{28}\right) = t \log 1.05 \qquad \text{Power Property}$$

$$t = \frac{\log\left(\frac{50}{28}\right)}{\log 1.05} \qquad \text{Divide each side by log 1.05.}$$

$$t \approx 11.9 \text{ years} \qquad \text{Use a calculator.}$$

If raises are given annually, she will be making over \$50,000 in 12 years.

Exercises

1. **BACTERIA GROWTH** A certain strain of bacteria grows from 40 to 326 in 120 minutes. Find k for the growth formula $y = ae^{kt}$, where t is in minutes.

2. **INVESTMENT** Carl plans to invest \$500 at 8.25% interest, compounded continuously. How long will it take for his money to triple?

3. **SCHOOL POPULATION** There are currently 850 students at the high school, which represents full capacity. The town plans an addition to house 400 more students. If the school population grows at 7.8% per year, in how many years will the new addition be full?

4. **EXERCISE** Hugo begins a walking program by walking $\frac{1}{2}$ mile per day for one week. Each week thereafter he increases his mileage by 10%. After how many weeks is he walking more than 5 miles per day?

5. **VOCABULARY GROWTH** When Emily was 18 months old, she had a 10-word vocabulary. By the time she was 5 years old (60 months), her vocabulary was 2500 words. If her vocabulary increased at a constant percent per month, what was that increase?

Study Guide and Intervention **124** Glencoe Algebra 2

Copyright © Glencoe/McGraw-Hill, a division of The McGraw-Hill Companies, Inc.

10-1 Study Guide and Intervention

Midpoint and Distance Formulas

The Midpoint Formula

Midpoint Formula	The midpoint M of a segment with endpoints (x_1, y_1) and (x_2, y_2) is $\left(\dfrac{x_1 + x_2}{2}, \dfrac{y_1 + y_2}{2}\right)$.

Example 1 Find the midpoint of the line segment with endpoints at $(4, -7)$ and $(-2, 3)$.

$$\left(\frac{x_1 + x_2}{2}, \frac{y_1 + y_2}{2}\right) = \left(\frac{4 + (-2)}{2}, \frac{-7 + 3}{2}\right)$$

$$= \left(\frac{2}{2}, \frac{-4}{2}\right) \text{ or } (1, -2)$$

The midpoint of the segment is $(1, -2)$.

Example 2 A diameter $\overline{AB}$ of a circle has endpoints $A(5, -11)$ and $B(-7, 6)$. What are the coordinates of the center of the circle?

The center of the circle is the midpoint of all of its diameters.

$$\left(\frac{x_1 + x_2}{2}, \frac{y_1 + y_2}{2}\right) = \left(\frac{5 + (-7)}{2}, \frac{-11 + 6}{2}\right)$$

$$= \left(\frac{-2}{2}, \frac{-5}{2}\right) \text{ or } \left(-1, -2\frac{1}{2}\right)$$

The circle has center $\left(-1, -2\frac{1}{2}\right)$.

Exercises

Find the midpoint of each line segment with endpoints at the given coordinates.

1. $(12, 7)$ and $(-2, 11)$

2. $(-8, -3)$ and $(10, 9)$

3. $(4, 15)$ and $(10, 1)$

4. $(-3, -3)$ and $(3, 3)$

5. $(15, 6)$ and $(12, 14)$

6. $(22, -8)$ and $(-10, 6)$

7. $(3, 5)$ and $(-6, 11)$

8. $(8, -15)$ and $(-7, 13)$

9. $(2.5, -6.1)$ and $(7.9, 13.7)$

10. $(-7, -6)$ and $(-1, 24)$

11. $(3, -10)$ and $(30, -20)$

12. $(-9, 1.7)$ and $(-11, 1.3)$

13. Segment $\overline{MN}$ has midpoint P. If M has coordinates $(14, -3)$ and P has coordinates $(-8, 6)$, what are the coordinates of N?

14. Circle R has a diameter $\overline{ST}$. If R has coordinates $(-4, -8)$ and S has coordinates $(1, 4)$, what are the coordinates of T?

15. Segment $\overline{AD}$ has midpoint B, and $\overline{BD}$ has midpoint C. If A has coordinates $(-5, 4)$ and C has coordinates $(10, 11)$, what are the coordinates of B and D?

10-1 Study Guide and Intervention (continued)

Midpoint and Distance Formulas

The Distance Formula

Distance Formula	The distance between two points (x_1, y_1) and (x_2, y_2) is given by $d = \sqrt{(x_2 - x_1)^2 + (y_2 - y_1)^2}$.

Example 1 What is the distance between $(8, -2)$ and $(-6, -8)$?

$$d = \sqrt{(x_2 - x_1)^2 + (y_2 - y_1)^2} \quad \text{Distance Formula}$$

$$= \sqrt{(-6 - 8)^2 + [-8 - (-2)]^2} \quad \text{Let } (x_1, y_1) = (8, -2) \text{ and } (x_2, y_2) = (-6, -8).$$

$$= \sqrt{(-14)^2 + (-6)^2} \quad \text{Subtract.}$$

$$= \sqrt{196 + 36} \text{ or } \sqrt{232} \quad \text{Simplify.}$$

The distance between the points is $\sqrt{232}$ or about 15.2 units.

Example 2 Find the perimeter and area of square $PQRS$ with vertices $P(-4, 1)$, $Q(-2, 7)$, $R(4, 5)$, and $S(2, -1)$.

Find the length of one side to find the perimeter and the area. Choose $\overline{PQ}$.

$$d = \sqrt{(x_2 - x_1)^2 + (y_2 - y_1)^2} \quad \text{Distance Formula}$$

$$= \sqrt{[-4 - (-2)]^2 + (1 - 7)^2} \quad \text{Let } (x_1, y_1) = (-4, 1) \text{ and } (x_2, y_2) = (-2, 7).$$

$$= \sqrt{(-2)^2 + (-6)^2} \quad \text{Subtract.}$$

$$= \sqrt{40} \text{ or } 2\sqrt{10} \quad \text{Simplify.}$$

Since one side of the square is $2\sqrt{10}$, the perimeter is $8\sqrt{10}$ units. The area is $(2\sqrt{10})^2$, or 40 units^2.

Exercises

Find the distance between each pair of points with the given coordinates.

1. $(3, 7)$ and $(-1, 4)$

2. $(-2, -10)$ and $(10, -5)$

3. $(6, -6)$ and $(-2, 0)$

4. $(7, 2)$ and $(4, -1)$

5. $(-5, -2)$ and $(3, 4)$

6. $(11, 5)$ and $(16, 9)$

7. $(-3, 4)$ and $(6, -11)$

8. $(13, 9)$ and $(11, 15)$

9. $(-15, -7)$ and $(2, 12)$

10. Rectangle $ABCD$ has vertices $A(1, 4)$, $B(3, 1)$, $C(-3, -2)$, and $D(-5, 1)$. Find the perimeter and area of $ABCD$.

11. Circle R has diameter $\overline{ST}$ with endpoints $S(4, 5)$ and $T(-2, -3)$. What are the circumference and area of the circle? (Express your answer in terms of π.)

10-2 Study Guide and Intervention

Parabolas

Equations of Parabolas A parabola is a curve consisting of all points in the coordinate plane that are the same distance from a given point (the **focus**) and a given line (the **directrix**). The following chart summarizes important information about parabolas.

Standard Form of Equation	$y = a(x - h)^2 + k$	$x = a(y - k)^2 + h$
Axis of Symmetry	$x = h$	$y = k$
Vertex	(h, k)	(h, k)
Focus	$\left(h, k + \dfrac{1}{4a}\right)$	$\left(h + \dfrac{1}{4a}, k\right)$
Directrix	$y = k - \dfrac{1}{4a}$	$x = h - \dfrac{1}{4a}$
Direction of Opening	upward if $a > 0$, downward if $a < 0$	right if $a > 0$, left if $a < 0$
Length of Latus Rectum	$\left\|\dfrac{1}{a}\right\|$ units	$\left\|\dfrac{1}{a}\right\|$ units

Example Identify the coordinates of the vertex and focus, the equations of the axis of symmetry and directrix, and the direction of opening of the parabola with equation $y = 2x^2 - 12x - 25$.

$y = 2x^2 - 12x - 25$	Original equation
$y = 2(x^2 - 6x) - 25$	Factor 2 from the *x*-terms.
$y = 2(x^2 - 6x + \blacksquare) - 25 - 2(\blacksquare)$	Complete the square on the right side.
$y = 2(x^2 - 6x + 9) - 25 - 2(9)$	The 9 added to complete the square is multiplied by 2.
$y = 2(x - 3)^2 - 43$	Write in standard form.

The vertex of this parabola is located at $(3, -43)$, the focus is located at $\left(3, -42\dfrac{7}{8}\right)$, the

equation of the axis of symmetry is $x = 3$, and the equation of the directrix is $y = -43\dfrac{1}{8}$. The parabola opens upward.

Exercises

Identify the coordinates of the vertex and focus, the equations of the axis of symmetry and directrix, and the direction of opening of the parabola with the given equation.

1. $y = x^2 + 6x - 4$ **2.** $y = 8x - 2x^2 + 10$ **3.** $x = y^2 - 8y + 6$

Write an equation of each parabola described below.

4. focus $(-2, 3)$, directrix $x = -2\dfrac{1}{12}$ **5.** vertex $(5, 1)$, focus $\left(4\dfrac{11}{12}, 1\right)$

10-2 **Study Guide and Intervention** *(continued)*

Parabolas

Graph Parabolas To graph an equation for a parabola, first put the given equation in standard form.

$y = a(x - h)^2 + k$ for a parabola opening up or down, or

$x = a(y - k)^2 + h$ for a parabola opening to the left or right

Use the values of a, h, and k to determine the vertex, focus, axis of symmetry, and length of the latus rectum. The vertex and the endpoints of the latus rectum give three points on the parabola. If you need more points to plot an accurate graph, substitute values for points near the vertex.

Example Graph $y = \frac{1}{3}(x - 1)^2 + 2$.

In the equation, $a = \frac{1}{3}$, $h = 1$, $k = 2$.

The parabola opens up, since $a > 0$.

vertex: $(1, 2)$

axis of symmetry: $x = 1$

focus: $\left(1, 2 + \dfrac{1}{4\left(\frac{1}{3}\right)}\right)$ or $\left(1, 2\frac{3}{4}\right)$

length of latus rectum: $\left|\dfrac{1}{\frac{1}{3}}\right|$ or 3 units

endpoints of latus rectum: $\left(2\frac{1}{2}, 2\frac{3}{4}\right), \left(-\frac{1}{2}, 2\frac{3}{4}\right)$

Exercises

The coordinates of the focus and the equation of the directrix of a parabola are given. Write an equation for each parabola and draw its graph.

1. $(3, 5), y = 1$

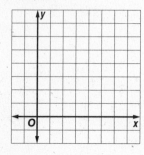

2. $(4, -4), y = -6$

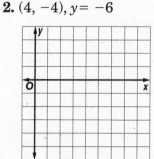

3. $(5, -1), x = 3$

10-3 Study Guide and Intervention

Circles

Equations of Circles The equation of a circle with center (h, k) and radius r units is $(x - h)^2 + (y - k)^2 = r^2$.

Example **Write an equation for a circle if the endpoints of a diameter are at $(-4, 5)$ and $(6, -3)$.**

Use the midpoint formula to find the center of the circle.

$$(h, k) = \left(\frac{x_1 + x_2}{2}, \frac{y_1 + y_2}{2}\right) \qquad \text{Midpoint formula}$$

$$= \left(\frac{-4 + 6}{2}, \frac{5 + (-3)}{2}\right) \qquad (x_1, y_1) = (-4, 5), (x_2, y_2) = (6, -3)$$

$$= \left(\frac{2}{2}, \frac{2}{2}\right) \text{ or } (1, 1) \qquad \text{Simplify.}$$

Use the coordinates of the center and one endpoint of the diameter to find the radius.

$$r = \sqrt{(x_2 - x_1)^2 + (y_2 - y_1)^2} \qquad \text{Distance formula}$$

$$r = \sqrt{(-4 - 1)^2 + (5 - 1)^2} \qquad (x_1, y_1) = (1, 1), (x_2, y_2) = (-4, 5)$$

$$= \sqrt{(-5)^2 + 4^2} = \sqrt{41} \qquad \text{Simplify.}$$

The radius of the circle is $\sqrt{41}$, so $r^2 = 41$.

An equation of the circle is $(x - 1)^2 + (y - 1)^2 = 41$.

Exercises

Write an equation for the circle that satisfies each set of conditions.

1. center $(8, -3)$, radius 6

2. center $(5, -6)$, radius 4

3. center $(-5, 2)$, passes through $(-9, 6)$

4. endpoints of a diameter at $(6, 6)$ and $(10, 12)$

5. center $(3, 6)$, tangent to the x-axis

6. center $(-4, -7)$, tangent to $x = 2$

7. center at $(-2, 8)$, tangent to $y = -4$

8. center $(7, 7)$, passes through $(12, 9)$

9. endpoints of a diameter are $(-4, -2)$ and $(8, 4)$

10. endpoints of a diameter are $(-4, 3)$ and $(6, -8)$

10-3 Study Guide and Intervention *(continued)*

Circles

Graph Circles To graph a circle, write the given equation in the standard form of the equation of a circle, $(x - h)^2 + (y - k)^2 = r^2$.

Plot the center (h, k) of the circle. Then use r to calculate and plot the four points $(h + r, k)$, $(h - r, k)$, $(h, k + r)$, and $(h, k - r)$, which are all points on the circle. Sketch the circle that goes through those four points.

Example Find the center and radius of the circle whose equation is $x^2 + 2x + y^2 + 4y = 11$. Then graph the circle.

$$x^2 + 2x + y^2 + 4y = 11$$
$$x^2 + 2x + \blacksquare + y^2 + 4y + \blacksquare = 11 + \blacksquare$$
$$x^2 + 2x + 1 + y^2 + 4y + 4 = 11 + 1 + 4$$
$$(x + 1)^2 + (y + 2)^2 = 16$$

Therefore, the circle has its center at $(-1, -2)$ and a radius of $\sqrt{16} = 4$. Four points on the circle are $(3, -2)$, $(-5, -2)$, $(-1, 2)$, and $(-1, -6)$.

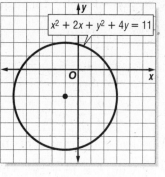

Exercises

Find the center and radius of the circle with the given equation. Then graph the circle.

1. $(x - 3)^2 + y^2 = 9$

2. $x^2 + (y + 5)^2 = 4$

3. $(x - 1)^2 + (y + 3)^2 = 9$

4. $(x - 2)^2 + (y + 4)^2 = 16$

5. $x^2 + y^2 - 10x + 8y + 16 = 0$

6. $x^2 + y^2 - 4x + 6y = 12$

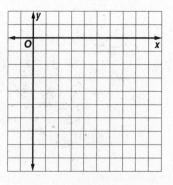

10-4 Study Guide and Intervention

Ellipses

Equations of Ellipses An **ellipse** is the set of all points in a plane such that the *sum* of the distances from two given points in the plane, called the foci, is constant. An ellipse has two axes of symmetry which contain the **major** and **minor axes**. In the table, the lengths a, b, and c are related by the formula $c^2 = a^2 - b^2$.

Standard Form of Equation	$\dfrac{(x-h)^2}{a^2} + \dfrac{(y-k)^2}{b^2} = 1$	$\dfrac{(y-k)^2}{a^2} + \dfrac{(x-h)^2}{b^2} = 1$
Center	(h, k)	(h, k)
Direction of Major Axis	Horizontal	Vertical
Foci	$(h+c, k), (h-c, k)$	$(h, k-c), (h, k+c)$
Length of Major Axis	$2a$ units	$2a$ units
Length of Minor Axis	$2b$ units	$2b$ units

Example **Write an equation for the ellipse shown.**

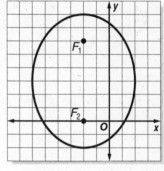

The length of the major axis is the distance between $(-2, -2)$ and $(-2, 8)$. This distance is 10 units.

$2a = 10$, so $a = 5$

The foci are located at $(-2, 6)$ and $(-2, 0)$, so $c = 3$.

$b^2 = a^2 - c^2$
$ = 25 - 9$
$ = 16$

The center of the ellipse is at $(-2, 3)$, so $h = -2$, $k = 3$, $a^2 = 25$, and $b^2 = 16$. The major axis is vertical.

An equation of the ellipse is $\dfrac{(y-3)^2}{25} + \dfrac{(x+2)^2}{16} = 1$.

Exercises

Write an equation for the ellipse that satisfies each set of conditions.

1. endpoints of major axis at $(-7, 2)$ and $(5, 2)$, endpoints of minor axis at $(-1, 0)$ and $(-1, 4)$

2. major axis 8 units long and parallel to the x-axis, minor axis 2 units long, center at $(-2, -5)$

3. endpoints of major axis at $(-8, 4)$ and $(4, 4)$, foci at $(-3, 4)$ and $(-1, 4)$

4. endpoints of major axis at $(3, 2)$ and $(3, -14)$, endpoints of minor axis at $(-1, -6)$ and $(7, -6)$

5. minor axis 6 units long and parallel to the x-axis, major axis 12 units long, center at $(6, 1)$

10-4 Study Guide and Intervention (continued)

Ellipses

Graph Ellipses To graph an ellipse, if necessary, write the given equation in the standard form of an equation for an ellipse.

$\dfrac{(x - h)^2}{a^2} + \dfrac{(y - k)^2}{b^2} = 1$ (for ellipse with major axis horizontal) or

$\dfrac{(y - k)^2}{a^2} + \dfrac{(x - h)^2}{b^2} = 1$ (for ellipse with major axis vertical)

Use the center (h, k) and the endpoints of the axes to plot four points of the ellipse. To make a more accurate graph, use a calculator to find some approximate values for x and y that satisfy the equation.

Example Graph the ellipse $4x^2 + 6y^2 + 8x - 36y = -34$.

$$4x^2 + 6y^2 + 8x - 36y = -34$$
$$4x^2 + 8x + 6y^2 - 36y = -34$$
$$4(x^2 + 2x + \blacksquare) + 6(y^2 - 6y + \blacksquare) = -34 + \blacksquare$$
$$4(x^2 + 2x + 1) + 6(y^2 - 6y + 9) = -34 + 58$$
$$4(x + 1)^2 + 6(y - 3)^2 = 24$$
$$\dfrac{(x + 1)^2}{6} + \dfrac{(y - 3)^2}{4} = 1$$

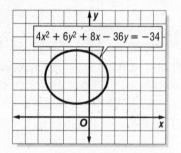

The center of the ellipse is $(-1, 3)$. Since $a^2 = 6$, $a = \sqrt{6}$. Since $b^2 = 4$, $b = 2$.

The length of the major axis is $2\sqrt{6}$, and the length of the minor axis is 4. Since the x-term has the greater denominator, the major axis is horizontal. Plot the endpoints of the axes. Then graph the ellipse.

Exercises

Find the coordinates of the center and the lengths of the major and minor axes for the ellipse with the given equation. Then graph the ellipse.

1. $\dfrac{y^2}{12} + \dfrac{x^2}{9} = 1$

2. $\dfrac{x^2}{25} + \dfrac{y^2}{4} = 1$

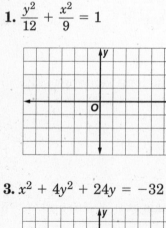

3. $x^2 + 4y^2 + 24y = -32$

4. $9x^2 + 6y^2 - 36x + 12y = 12$

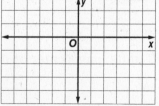

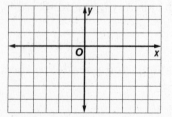

10-5 Study Guide and Intervention
Hyperbolas

Equations of Hyperbolas A **hyperbola** is the set of all points in a plane such that the absolute value of the *difference* of the distances from any point on the hyperbola to any two given points in the plane, called the **foci**, is constant.

In the table, the lengths a, b, and c are related by the formula $c^2 = a^2 + b^2$.

Standard Form of Equation	$\frac{(x-h)^2}{a^2} - \frac{(y-k)^2}{b^2} = 1$	$\frac{(y-k)^2}{a^2} - \frac{(x-h)^2}{b^2} = 1$
Equations of the Asymptotes	$y - k = \pm\frac{b}{a}(x-h)$	$y - k = \pm\frac{a}{b}(x-h)$
Transverse Axis	Horizontal	Vertical
Foci	$(h-c, k), (h+c, k)$	$(h, k-c), (h, k+c)$
Vertices	$(h-a, k), (h+a, k)$	$(h, k-a), (h, k+a)$

Example Write an equation for the hyperbola with vertices $(-2, 1)$ and $(6, 1)$ and foci $(-4, 1)$ and $(8, 1)$.

Use a sketch to orient the hyperbola correctly. The center of the hyperbola is the midpoint of the segment joining the two vertices. The center is $(\frac{-2+6}{2}, 1)$, or $(2, 1)$. The value of a is the distance from the center to a vertex, so $a = 4$. The value of c is the distance from the center to a focus, so $c = 6$.

$c^2 = a^2 + b^2$
$6^2 = 4^2 + b^2$
$b^2 = 36 - 16 = 20$

Use h, k, a^2, and b^2 to write an equation of the hyperbola.
$\frac{(x-2)^2}{16} - \frac{(y-1)^2}{20} = 1$

Exercises

Write an equation for the hyperbola that satisfies each set of conditions.

1. vertices $(-7, 0)$ and $(7, 0)$, conjugate axis of length 10

2. vertices $(-2, -3)$ and $(4, -3)$, foci $(-5, -3)$ and $(7, -3)$

3. vertices $(4, 3)$ and $(4, -5)$, conjugate axis of length 4

4. vertices $(-8, 0)$ and $(8, 0)$, equation of asymptotes $y = \pm\frac{1}{6}x$

5. vertices $(-4, 6)$ and $(-4, -2)$, foci $(-4, 10)$ and $(-4, -6)$

10-5 Study Guide and Intervention (continued)

Hyperbolas

Graph Hyperbolas To graph a hyperbola, write the given equation in the standard form of an equation for a hyperbola

$$\frac{(x-h)^2}{a^2} - \frac{(y-k)^2}{b^2} = 1 \text{ if the branches of the hyperbola open left and right, or}$$

$$\frac{(y-k)^2}{a^2} - \frac{(x-h)^2}{b^2} = 1 \text{ if the branches of the hyperbola open up and down}$$

Graph the point (h, k), which is the center of the hyperbola. Draw a rectangle with dimensions $2a$ and $2b$ and center (h, k). If the hyperbola opens left and right, the vertices are $(h - a, k)$ and $(h + a, k)$. If the hyperbola opens up and down, the vertices are $(h, k - a)$ and $(h, k + a)$.

Example **Draw the graph of $6y^2 - 4x^2 - 36y - 8x = -26$.**

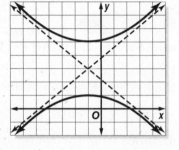

Complete the squares to get the equation in standard form.
$6y^2 - 4x^2 - 36y - 8x = -26$
$6(y^2 - 6y + \blacksquare) - 4(x^2 + 2x + \blacksquare) = -26 + \blacksquare$
$6(y^2 - 6y + 9) - 4(x^2 + 2x + 1) = -26 + 50$
$6(y - 3)^2 - 4(x + 1)^2 = 24$
$\dfrac{(y-3)^2}{4} - \dfrac{(x+1)^2}{6} = 1$

The center of the hyperbola is $(-1, 3)$.
According to the equation, $a^2 = 4$ and $b^2 = 6$, so $a = 2$ and $b = \sqrt{6}$.
The transverse axis is vertical, so the vertices are $(-1, 5)$ and $(-1, 1)$. Draw a rectangle with vertical dimension 4 and horizontal dimension $2\sqrt{6} \approx 4.9$. The diagonals of this rectangle are the asymptotes. The branches of the hyperbola open up and down. Use the vertices and the asymptotes to sketch the hyperbola.

Exercises

Find the coordinates of the vertices and foci and the equations of the asymptotes for the hyperbola with the given equation. Then graph the hyperbola.

1. $\dfrac{x^2}{4} - \dfrac{y^2}{16} = 1$ 2. $(y - 3)^2 - \dfrac{(x + 2)^2}{9} = 1$ 3. $\dfrac{y^2}{16} - \dfrac{x^2}{9} = 1$

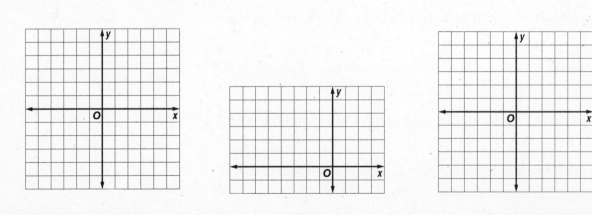

10-6 Study Guide and Intervention

Conic Sections

Standard Form Any conic section in the coordinate plane can be described by an equation of the form

$$Ax^2 + Bxy + Cy^2 + Dx + Ey + F = 0, \text{ where } A, B, \text{ and } C \text{ are not all zero.}$$

One way to tell what kind of conic section an equation represents is to rearrange terms and complete the square, if necessary, to get one of the standard forms from an earlier lesson. This method is especially useful if you are going to graph the equation.

Example **Write the equation $3x^2 - 4y^2 - 30x - 8y + 59 = 0$ in standard form. State whether the graph of the equation is a *parabola*, *circle*, *ellipse*, or *hyperbola*.**

$3x^2 - 4y^2 - 30x - 8y + 59 = 0$	Original equation
$3x^2 - 30x - 4y^2 - 8y = -59$	Isolate terms.
$3(x^2 - 10x + \blacksquare) - 4(y^2 + 2y + \blacksquare) = -59 + \blacksquare + \blacksquare$	Factor out common multiples.
$3(x^2 - 10x + 25) - 4(y^2 + 2y + 1) = -59 + 3(25) + (-4)(1)$	Complete the squares.
$3(x - 5)^2 - 4(y + 1)^2 = 12$	Simplify.
$\dfrac{(x - 5)^2}{4} - \dfrac{(y + 1)^2}{3} = 1$	Divide each side by 12.

The graph of the equation is a hyperbola with its center at $(5, -1)$. The length of the transverse axis is 4 units and the length of the conjugate axis is $2\sqrt{3}$ units.

Exercises

Write each equation in standard form. State whether the graph of the equation is a *parabola*, *circle*, *ellipse*, or *hyperbola*.

1. $x^2 + y^2 - 6x + 4y + 3 = 0$

2. $x^2 + 2y^2 + 6x - 20y + 53 = 0$

3. $6x^2 - 60x - y + 161 = 0$

4. $x^2 + y^2 - 4x - 14y + 29 = 0$

5. $6x^2 - 5y^2 + 24x + 20y - 56 = 0$

6. $3y^2 + x - 24y + 46 = 0$

7. $x^2 - 4y^2 - 16x + 24y - 36 = 0$

8. $x^2 + 2y^2 + 8x + 4y + 2 = 0$

9. $4x^2 + 48x + y + 158 = 0$

10. $3x^2 + y^2 - 48x - 4y + 184 = 0$

11. $-3x^2 + 2y^2 - 18x + 20y + 5 = 0$

12. $x^2 + y^2 + 8x + 2y + 8 = 0$

10-6 Study Guide and Intervention (continued)

Conic Sections

Identify Conic Sections If you are given an equation of the form
$$Ax^2 + Bxy + Cy^2 + Dx + Ey + F = 0, \text{ with } B = 0,$$
you can determine the type of conic section just by considering the values of A and C. Refer to the following chart.

Relationship of A and C	Type of Conic Section
$A = 0$ or $C = 0$, but not both.	parabola
$A = C$	circle
A and C have the same sign, but $A \neq C$.	ellipse
A and C have opposite signs.	hyperbola

Example **Without writing the equation in standard form, state whether the graph of each equation is a *parabola, circle, ellipse,* or *hyperbola*.**

a. $3x^2 - 3y^2 + 5x + 12 = 0$

$A = 3$ and $C = -3$ have opposite signs, so the graph of the equation is a hyperbola.

b. $y^2 = 7y - 2x + 13$

$A = 0$, so the graph of the equation is a parabola.

Exercises

Without writing the equation in standard form, state whether the graph of each equation is a *parabola, circle, ellipse,* or *hyperbola*.

1. $x^2 = 17x - 5y + 8$

2. $2x^2 + 2y^2 - 3x + 4y = 5$

3. $4x^2 - 8x = 4y^2 - 6y + 10$

4. $8(x - x^2) = 4(2y^2 - y) - 100$

5. $6y^2 - 18 = 24 - 4x^2$

6. $y = 27x - y^2$

7. $x^2 = 4(y - y^2) + 2x - 1$

8. $10x - x^2 - 2y^2 = 5y$

9. $x = y^2 - 5y + x^2 - 5$

10. $11x^2 - 7y^2 = 77$

11. $3x^2 + 4y^2 = 50 + y^2$

12. $y^2 = 8x - 11$

13. $9y^2 - 99y = 3(3x - 3x^2)$

14. $6x^2 - 4 = 5y^2 - 3$

15. $111 = 11x^2 + 10y^2$

16. $120x^2 - 119y^2 + 118x - 117y = 0$

17. $3x^2 = 4y^2 + 12$

18. $150 - x^2 = 120 - y$

10-7 Study Guide and Intervention

Solving Quadratic Systems

Systems of Quadratic Equations Like systems of linear equations, systems of quadratic equations can be solved by substitution and elimination. If the graphs are a conic section and a line, the system will have 0, 1, or 2 solutions. If the graphs are two conic sections, the system will have 0, 1, 2, 3, or 4 solutions.

Example Solve the system of equations. $y = x^2 - 2x - 15$
$$x + y = -3$$

Rewrite the second equation as $y = -x - 3$ and substitute into the first equation.

$-x - 3 = x^2 - 2x - 15$
$\quad 0 = x^2 - x - 12 \quad$ Add $x + 3$ to each side.
$\quad 0 = (x - 4)(x + 3) \quad$ Factor.

Use the Zero Product property to get
$x = 4$ or $x = -3$.

Substitute these values for x in $x + y = -3$:

$4 + y = -3$ or $-3 + y = -3$
$\quad y = -7 \qquad\qquad y = 0$

The solutions are $(4, -7)$ and $(-3, 0)$.

Exercises

Find the exact solution(s) of each system of equations.

1. $y = x^2 - 5$
$\quad y = x - 3$

2. $x^2 + (y - 5)^2 = 25$
$\quad y = -x^2$

3. $x^2 + (y - 5)^2 = 25$
$\quad y = x^2$

4. $x^2 + y^2 = 9$
$\quad x^2 + y = 3$

5. $x^2 - y^2 = 1$
$\quad x^2 + y^2 = 16$

6. $y = x - 3$
$\quad x = y^2 - 4$

10-7 Study Guide and Intervention *(continued)*

Solving Quadratic Systems

Systems of Quadratic Inequalities Systems of quadratic inequalities can be solved by graphing.

Example 1 Solve the system of inequalities by graphing.

$x^2 + y^2 \le 25$

$\left(x - \dfrac{5}{2}\right)^2 + y^2 \ge \dfrac{25}{4}$

The graph of $x^2 + y^2 \le 25$ consists of all points on or inside the circle with center $(0, 0)$ and radius 5. The graph of $\left(x - \dfrac{5}{2}\right)^2 + y^2 \ge \dfrac{25}{4}$ consists of all points on or outside the circle with center $\left(\dfrac{5}{2}, 0\right)$ and radius $\dfrac{5}{2}$. The solution of the system is the set of points in both regions.

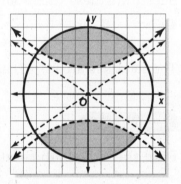

Example 2 Solve the system of inequalities by graphing.

$x^2 + y^2 \le 25$

$\dfrac{y^2}{4} - \dfrac{x^2}{9} > 1$

The graph of $x^2 + y^2 \le 25$ consists of all points on or inside the circle with center $(0, 0)$ and radius 5. The graph of $\dfrac{y^2}{4} - \dfrac{x^2}{9} > 1$ are the points "inside" but not on the branches of the hyperbola shown. The solution of the system is the set of points in both regions.

Exercises

Solve each system of inequalities below by graphing.

1. $\dfrac{x^2}{16} + \dfrac{y^2}{4} \le 1$

    $y > \dfrac{1}{2}x - 2$

2. $x^2 + y^2 \le 169$

    $x^2 + 9y^2 \ge 225$

3. $y \ge (x - 2)^2$

    $(x + 1)^2 + (y + 1)^2 \le 16$

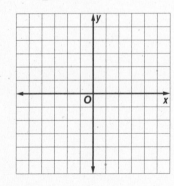
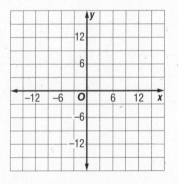
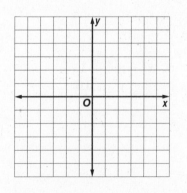

NAME _____ DATE _____ PERIOD _____

11-1 Study Guide and Intervention

Arithmetic Sequences

Arithmetic Sequences An **arithmetic sequence** is a sequence of numbers in which each **term** after the first term is found by adding the **common difference** to the preceding term.

nth Term of an Arithmetic Sequence	$a_n = a_1 + (n - 1)d$, where a_1 is the first term, d is the common difference, and n is any positive integer

Example 1 Find the next four terms of the arithmetic sequence 7, 11, 15, ...

Find the common difference by subtracting two consecutive terms.

$11 - 7 = 4$ and $15 - 11 = 4$, so $d = 4$.

Now add 4 to the third term of the sequence, and then continue adding 4 until the four terms are found. The next four terms of the sequence are 19, 23, 27, and 31.

Example 2 Find the thirteenth term of the arithmetic sequence with $a_1 = 21$ and $d = -6$.

Use the formula for the nth term of an arithmetic sequence with $a_1 = 21$, $n = 13$, and $d = -6$.

$a_n = a_1 + (n - 1)d$ Formula for nth term

$a_{13} = 21 + (13 - 1)(-6)$ $n = 13$, $a_1 = 21$, $d = -6$

$a_{13} = -51$ Simplify.

The thirteenth term is -51.

Example 3 Write an equation for the nth term of the arithmetic sequence $-14, -5, 4, 13, \ldots$.

In this sequence $a_1 = -14$ and $d = 9$. Use the formula for a_n to write an equation.

$a_n = a_1 + (n - 1)d$ Formula for the nth term

   $= -14 + (n - 1)9$ $a_1 = -14$, $d = 9$

   $= -14 + 9n - 9$ Distributive Property

   $= 9n - 23$ Simplify.

Exercises

Find the next four terms of each arithmetic sequence.

1. 106, 111, 116, ...   **2.** $-28, -31, -34, \ldots$ **3.** 207, 194, 181, ...

Find the first five terms of each arithmetic sequence described.

4. $a_1 = 101$, $d = 9$ **5.** $a_1 = -60$, $d = 4$ **6.** $a_1 = 210$, $d = -40$

Find the indicated term of each arithmetic sequence.

7. $a_1 = 4$, $d = 6$, $n = 14$ **8.** $a_1 = -4$, $d = -2$, $n = 12$

9. $a_1 = 80$, $d = -8$, $n = 21$ **10.** a_{10} for $0, -3, -6, -9, \ldots$

Write an equation for the nth term of each arithmetic sequence.

11. 18, 25, 32, 39, ...   **12.** $-110, -85, -60, -35, \ldots$ **13.** 6.2, 8.1, 10.0, 11.9, ...

Study Guide and Intervention   **139**   Glencoe Algebra 2

Copyright © Glencoe/McGraw-Hill, a division of The McGraw-Hill Companies, Inc.

11-1 Study Guide and Intervention *(continued)*

Arithmetic Sequences

Arithmetic Means The **arithmetic means** of an arithmetic sequence are the terms between any two nonsuccessive terms of the sequence.

To find the k arithmetic means between two terms of a sequence, use the following steps.

Step 1 Let the two terms given be a_1 and a_n, where $n = k + 2$.
Step 2 Substitute in the formula $a_n = a_1 + (n - 1)d$.
Step 3 Solve for d, and use that value to find the k arithmetic means:
$\qquad a_1 + d, a_1 + 2d, \ldots, a_1 + kd$.

Example   **Find the five arithmetic means between 37 and 121.**

You can use the nth term formula to find the common difference. In the sequence, $37, \underline{\ ?\ }, \underline{\ ?\ }, \underline{\ ?\ }, \underline{\ ?\ }, \underline{\ ?\ }, 121, \ldots, a_1$ is 37 and a_7 is 121.

$$a_n = a_1 + (n - 1)d \qquad \text{Formula for the } n\text{th term}$$
$$121 = 37 + (7 - 1)d \qquad a_1 = 37, a_7 = 121, n = 7$$
$$121 = 37 + 6d \qquad \text{Simplify.}$$
$$84 = 6d \qquad \text{Subtract 37 from each side.}$$
$$d = 14 \qquad \text{Divide each side by 6.}$$

Now use the value of d to find the five arithmetic means.

$$37 \searrow 51 \searrow 65 \searrow 79 \searrow 93 \searrow 107 \searrow 121$$
$$+14 \quad +14 \quad +14 \quad +14 \quad +14 \quad +14$$

The arithmetic means are 51, 65, 79, 93, and 107.

Exercises

Find the arithmetic means in each sequence.

1. $5, \underline{\ ?\ }, \underline{\ ?\ }, \underline{\ ?\ }, -3$
    2. $18, \underline{\ ?\ }, \underline{\ ?\ }, \underline{\ ?\ }, -2$
    3. $16, \underline{\ ?\ }, \underline{\ ?\ }, 37$

4. $108, \underline{\ ?\ }, \underline{\ ?\ }, \underline{\ ?\ }, \underline{\ ?\ }, 48$
    5. $-14, \underline{\ ?\ }, \underline{\ ?\ }, \underline{\ ?\ }, -30$
    6. $29, \underline{\ ?\ }, \underline{\ ?\ }, \underline{\ ?\ }, 89$

7. $61, \underline{\ ?\ }, \underline{\ ?\ }, \underline{\ ?\ }, \underline{\ ?\ }, 116$
         8. $45, \underline{\ ?\ }, \underline{\ ?\ }, \underline{\ ?\ }, \underline{\ ?\ }, \underline{\ ?\ }, 81$

9. $-18, \underline{\ ?\ }, \underline{\ ?\ }, \underline{\ ?\ }, 14$
       10. $-40, \underline{\ ?\ }, \underline{\ ?\ }, \underline{\ ?\ }, \underline{\ ?\ }, \underline{\ ?\ }, -82$

11. $100, \underline{\ ?\ }, \underline{\ ?\ }, 235$
        12. $80, \underline{\ ?\ }, \underline{\ ?\ }, \underline{\ ?\ }, \underline{\ ?\ }, -30$

13. $450, \underline{\ ?\ }, \underline{\ ?\ }, \underline{\ ?\ }, 570$
     14. $27, \underline{\ ?\ }, \underline{\ ?\ }, \underline{\ ?\ }, \underline{\ ?\ }, \underline{\ ?\ }, 57$

15. $125, \underline{\ ?\ }, \underline{\ ?\ }, \underline{\ ?\ }, 185$
     16. $230, \underline{\ ?\ }, \underline{\ ?\ }, \underline{\ ?\ }, \underline{\ ?\ }, \underline{\ ?\ }, 128$

17. $-20, \underline{\ ?\ }, \underline{\ ?\ }, \underline{\ ?\ }, \underline{\ ?\ }, 370$
    18. $48, \underline{\ ?\ }, \underline{\ ?\ }, \underline{\ ?\ }, 100$

11-2 Study Guide and Intervention

Arithmetic Series

Arithmetic Series An **arithmetic series** is the sum of consecutive terms of an arithmetic sequence.

Sum of an Arithmetic Series	The sum S_n of the first n terms of an arithmetic series is given by the formula $S_n = \frac{n}{2}[2a_1 + (n-1)d]$ or $S_n = \frac{n}{2}(a_1 + a_n)$

Example 1 Find S_n for the arithmetic series with $a_1 = 14$, $a_n = 101$, and $n = 30$.

Use the sum formula for an arithmetic series.

$$S_n = \frac{n}{2}(a_1 + a_n) \qquad \text{Sum formula}$$

$$S_{30} = \frac{30}{2}(14 + 101) \qquad n = 30, a_1 = 14, a_n = 101$$

$$= 15(115) \qquad \text{Simplify.}$$

$$= 1725 \qquad \text{Multiply.}$$

The sum of the series is 1725.

Example 2 Find the sum of all positive odd integers less than 180.

The series is $1 + 3 + 5 + \ldots + 179$.

Find n using the formula for the nth term of an arithmetic sequence.

$$a_n = a_1 + (n-1)d \qquad \text{Formula for } n\text{th term}$$

$$179 = 1 + (n-1)2 \qquad a_n = 179, a_1 = 1, d = 2$$

$$179 = 2n - 1 \qquad \text{Simplify.}$$

$$180 = 2n \qquad \text{Add 1 to each side.}$$

$$n = 90 \qquad \text{Divide each side by 2.}$$

Then use the sum formula for an arithmetic series.

$$S_n = \frac{n}{2}(a_1 + a_n) \qquad \text{Sum formula}$$

$$S_{90} = \frac{90}{2}(1 + 179) \qquad n = 90, a_1 = 1, a_n = 179$$

$$= 45(180) \qquad \text{Simplify.}$$

$$= 8100 \qquad \text{Multiply.}$$

The sum of all positive odd integers less than 180 is 8100.

Exercises

Find S_n for each arithmetic series described.

1. $a_1 = 12, a_n = 100,$ $n = 12$

2. $a_1 = 50, a_n = -50,$ $n = 15$

3. $a_1 = 60, a_n = -136,$ $n = 50$

4. $a_1 = 20, d = 4,$ $a_n = 112$

5. $a_1 = 180, d = -8,$ $a_n = 68$

6. $a_1 = -8, d = -7,$ $a_n = -71$

7. $a_1 = 42, n = 8, d = 6$

8. $a_1 = 4, n = 20, d = 2\frac{1}{2}$

9. $a_1 = 32, n = 27, d = 3$

Find the sum of each arithmetic series.

10. $8 + 6 + 4 + \ldots + -10$

11. $16 + 22 + 28 + \ldots + 112$

Find the first three terms of each arithmetic series described.

13. $a_1 = 12, a_n = 174,$ $S_n = 1767$

14. $a_1 = 80, a_n = -115,$ $S_n = -245$

15. $a_1 = 6.2, a_n = 12.6,$ $S_n = 84.6$

11-2 Study Guide and Intervention *(continued)*

Arithmetic Series

Sigma Notation A shorthand notation for representing a series makes use of the Greek letter Σ. The **sigma notation** for the series $6 + 12 + 18 + 24 + 30$ is $\displaystyle\sum_{n=1}^{5} 6n$.

Example Evaluate $\displaystyle\sum_{k=1}^{18} (3k + 4)$.

The sum is an arithmetic series with common difference 3. Substituting $k = 1$ and $k = 18$ into the expression $3k + 4$ gives $a_1 = 3(1) + 4 = 7$ and $a_{18} = 3(18) + 4 = 58$. There are 18 terms in the series, so $n = 18$. Use the formula for the sum of an arithmetic series.

$S_n = \dfrac{n}{2}(a_1 + a_n)$ Sum formula

$S_{18} = \dfrac{18}{2}(7 + 58)$ $n = 18$, $a_1 = 7$, $a_n = 58$

$\quad = 9(65)$ Simplify.

$\quad = 585$ Multiply.

So $\displaystyle\sum_{k=1}^{18} (3k + 4) = 585$.

Exercises

Find the sum of each arithmetic series.

1. $\displaystyle\sum_{n=1}^{20} (2n + 1)$ 2. $\displaystyle\sum_{n=5}^{25} (x - 1)$ 3. $\displaystyle\sum_{k=1}^{18} (2k - 7)$

4. $\displaystyle\sum_{r=10}^{75} (2r - 200)$ 5. $\displaystyle\sum_{x=1}^{15} (6x + 3)$ 6. $\displaystyle\sum_{t=1}^{50} (500 - 6t)$

7. $\displaystyle\sum_{k=1}^{80} (100 - k)$ 8. $\displaystyle\sum_{n=20}^{85} (n - 100)$ 9. $\displaystyle\sum_{s=1}^{200} 3s$

10. $\displaystyle\sum_{m=14}^{28} (2m - 50)$ 11. $\displaystyle\sum_{p=1}^{36} (5p - 20)$ 12. $\displaystyle\sum_{j=12}^{32} (25 - 2j)$

13. $\displaystyle\sum_{n=18}^{42} (4n - 9)$ 14. $\displaystyle\sum_{n=20}^{50} (3n + 4)$ 15. $\displaystyle\sum_{j=5}^{44} (7j - 3)$

11-3 Study Guide and Intervention

Geometric Sequences

Geometric Sequences A **geometric sequence** is a sequence in which each term after the first is the product of the previous term and a constant called the **constant ratio**.

nth Term of a Geometric Sequence	$a_n = a_1 \cdot r^{n-1}$, where a_1 is the first term, r is the common ratio, and n is any positive integer

Example 1 Find the next two terms of the geometric sequence 1200, 480, 192, ….

Since $\dfrac{480}{1200} = 0.4$ and $\dfrac{192}{480} = 0.4$, the sequence has a common ratio of 0.4. The next two terms in the sequence are $192(0.4) = 76.8$ and $76.8(0.4) = 30.72$.

Example 2 Write an equation for the nth term of the geometric sequence 3.6, 10.8, 32.4, ….

In this sequence $a_1 = 3.6$ and $r = 3$. Use the nth term formula to write an equation.

$a_n = a_1 \cdot r^{n-1}$ Formula for nth term
$\quad = 3.6 \cdot 3^{n-1}$ $a_1 = 3.6, r = 3$

An equation for the nth term is $a_n = 3.6 \cdot 3^{n-1}$.

Exercises

Find the next two terms of each geometric sequence.

1. 6, 12, 24, …

2. 180, 60, 20, …

3. 2000, −1000, 500, …

4. 0.8, −2.4, 7.2, …

5. 80, 60, 45, …

6. 3, 16.5, 90.75, …

Find the first five terms of each geometric sequence described.

7. $a_1 = \dfrac{1}{9}, r = 3$

8. $a_1 = 240, r = -\dfrac{3}{4}$

9. $a_1 = 10, r = \dfrac{5}{2}$

Find the indicated term of each geometric sequence.

10. $a_1 = -10, r = 4, n = 2$

11. $a_1 = -6, r = -\dfrac{1}{2}, n = 8$

12. $a_3 = 9, r = -3, n = 7$

13. $a_4 = 16, r = 2, n = 10$

14. $a_4 = -54, r = -3, n = 6$

15. $a_1 = 8, r = \dfrac{2}{3}, n = 5$

Write an equation for the nth term of each geometric sequence.

16. 500, 350, 245, …

17. 8, 32, 128, …

18. 11, −24.2, 53.24, …

11-3 Study Guide and Intervention (continued)

Geometric Sequences

Geometric Means The **geometric means** of a geometric sequence are the terms between any two nonsuccessive terms of the sequence.

To find the k geometric means between two terms of a sequence, use the following steps.

Step 1	Let the two terms given be a_1 and a_n, where $n = k + 2$.
Step 2	Substitute in the formula $a_n = a_1 \cdot r^{n-1}$ $(= a_1 \cdot r^{k+1})$.
Step 3	Solve for r, and use that value to find the k geometric means: $a_1 \cdot r, a_1 \cdot r^2, \ldots, a_1 \cdot r^k$

Example Find the three geometric means between 8 and 40.5.

Use the nth term formula to find the value of r. In the sequence 8, __?__, __?__, __?__, 40.5, a_1 is 8 and a_5 is 40.5.

$$a_n = a_1 \cdot r^{n-1} \quad \text{Formula for } n\text{th term}$$
$$40.5 = 8 \cdot r^{5-1} \quad n = 5, a_1 = 8, a_5 = 40.5$$
$$5.0625 = r^4 \quad \text{Divide each side by 8.}$$
$$r = \pm 1.5 \quad \text{Take the fourth root of each side.}$$

There are two possible common ratios, so there are two possible sets of geometric means. Use each value of r to find the geometric means.

$r = 1.5$	$r = -1.5$
$a_2 = 8(1.5)$ or 12	$a_2 = 8(-1.5)$ or -12
$a_3 = 12(1.5)$ or 18	$a_3 = -12(-1.5)$ or 18
$a_4 = 18(1.5)$ or 27	$a_4 = 18(-1.5)$ or -27

The geometric means are 12, 18, and 27, or -12, 18, and -27.

Exercises

Find the geometric means in each sequence.

1. 5, __?__, __?__, __?__, 405

2. 5, __?__, __?__, 20.48

3. $\frac{3}{5}$, __?__, __?__, __?__, 375

4. -24, __?__, __?__, $\frac{1}{9}$

5. 12, __?__, __?__, __?__, __?__, __?__, $\frac{3}{16}$

6. 200, __?__, __?__, __?__, 414.72

7. $\frac{35}{49}$, __?__, __?__, __?__, __?__, $-12,005$

8. 4, __?__, __?__, __?__, $156\frac{1}{4}$

9. $-\frac{1}{81}$, __?__, __?__, __?__, __?__, __?__, -9

10. 100, __?__, __?__, __?__, 384.16

144

11-4 Study Guide and Intervention

Geometric Series

Geometric Series A **geometric series** is the indicated sum of consecutive terms of a geometric sequence.

Sum of a Geometric Series	The sum S_n of the first n terms of a geometric series is given by $S_n = \dfrac{a_1(1 - r^n)}{1 - r}$ or $S_n = \dfrac{a_1 - a_1 r^n}{1 - r}$, where $r \neq 1$.

Example 1 Find the sum of the first four terms of the geometric sequence for which $a_1 = 120$ and $r = \dfrac{1}{3}$.

$S_n = \dfrac{a_1(1 - r^n)}{1 - r}$ Sum formula

$S_4 = \dfrac{120\left(1 - \left(\frac{1}{3}\right)^4\right)}{1 - \frac{1}{3}}$ $n = 4, a_1 = 120, r = \dfrac{1}{3}$

≈ 177.78 Use a calculator.

The sum of the series is 177.78.

Example 2 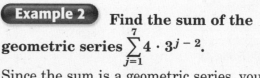 Find the sum of the geometric series $\displaystyle\sum_{j=1}^{7} 4 \cdot 3^{j-2}$.

Since the sum is a geometric series, you can use the sum formula.

$S_n = \dfrac{a_1(1 - r^n)}{1 - r}$ Sum formula

$S_7 = \dfrac{\frac{4}{3}(1 - 3^7)}{1 - 3}$ $n = 7, a_1 = \dfrac{4}{3}, r = 3$

≈ 1457.33 Use a calculator.

The sum of the series is 1457.33.

Exercises

Find S_n for each geometric series described.

1. $a_1 = 2, a_n = 486, r = 3$

2. $a_1 = 1200, a_n = 75, r = \dfrac{1}{2}$

3. $a_1 = \dfrac{1}{25}, a_n = 125, r = 5$

4. $a_1 = 3, r = \dfrac{1}{3}, n = 4$

5. $a_1 = 2, r = 6, n = 4$

6. $a_1 = 2, r = 4, n = 6$

7. $a_1 = 100, r = -\dfrac{1}{2}, n = 5$

8. $a_3 = 20, a_6 = 160, n = 8$

9. $a_4 = 16, a_7 = 1024, n = 10$

Find the sum of each geometric series.

10. $6 + 18 + 54 + \ldots$ to 6 terms

11. $\dfrac{1}{4} + \dfrac{1}{2} + 1 + \ldots$ to 10 terms

12. $\displaystyle\sum_{j=4}^{8} 2^j$

13. $\displaystyle\sum_{k=1}^{7} 3 \cdot 2^{k-1}$

11-4 Study Guide and Intervention *(continued)*

Geometric Series

Specific Terms You can use one of the formulas for the sum of a geometric series to help find a particular term of the series.

Example 1 Find a_1 in a geometric series for which $S_6 = 441$ and $r = 2$.

$$S_n = \frac{a_1(1 - r^n)}{1 - r} \quad \text{Sum formula}$$

$$441 = \frac{a_1(1 - 2^6)}{1 - 2} \quad S_6 = 441, r = 2, n = 6$$

$$441 = \frac{-63a_1}{-1} \quad \text{Subtract.}$$

$$a_1 = \frac{441}{63} \quad \text{Divide.}$$

$$a_1 = 7 \quad \text{Simplify.}$$

The first term of the series is 7.

Example 2 Find a_1 in a geometric series for which $S_n = 244$, $a_n = 324$, and $r = -3$.

Since you do not know the value of n, use the alternate sum formula.

$$S_n = \frac{a_1 - a_n r}{1 - r} \quad \text{Alternate sum formula}$$

$$244 = \frac{a_1 - (324)(-3)}{1 - (-3)} \quad S_n = 244, a_n = 324, r = -3$$

$$244 = \frac{a_1 + 972}{4} \quad \text{Simplify.}$$

$$976 = a_1 + 972 \quad \text{Multiply each side by 4.}$$

$$a_1 = 4 \quad \text{Subtract 972 from each side.}$$

The first term of the series is 4.

Example 3 Find a_4 in a geometric series for which $S_n = 796.875$, $r = \frac{1}{2}$, and $n = 8$.

First use the sum formula to find a_1.

$$S_n = \frac{a_1(1 - r^n)}{1 - r} \quad \text{Sum formula}$$

$$796.875 = \frac{a_1\left(1 - \left(\frac{1}{2}\right)^8\right)}{1 - \frac{1}{2}} \quad S_8 = 796.875, r = \frac{1}{2}, n = 8$$

$$796.875 = \frac{0.99609375a_1}{0.5} \quad \text{Use a calculator.}$$

$$a_1 = 400$$

Since $a_4 = a_1 \cdot r^3$, $a_4 = 400\left(\frac{1}{2}\right)^3 = 50$. The fourth term of the series is 50.

Exercises

Find the indicated term for each geometric series described.

1. $S_n = 726$, $a_n = 486$, $r = 3$; a_1

2. $S_n = 850$, $a_n = 1280$, $r = -2$; a_1

3. $S_n = 1023.75$, $a_n = 512$, $r = 2$; a_1

4. $S_n = 118.125$, $a_n = -5.625$, $r = -\frac{1}{2}$; a_1

5. $S_n = 183$, $r = -3$, $n = 5$; a_1

6. $S_n = 1705$, $r = 4$, $n = 5$; a_1

7. $S_n = 52{,}084$, $r = -5$, $n = 7$; a_1

8. $S_n = 43{,}690$, $r = \frac{1}{4}$, $n = 8$; a_1

9. $S_n = 381$, $r = 2$, $n = 7$; a_4

11-5 Study Guide and Intervention

Infinite Geometric Series

Infinite Geometric Series A geometric series that does not end is called an **infinite geometric series**. Some infinite geometric series have sums, but others do not because the **partial sums** increase without approaching a limiting value.

Sum of an Infinite Geometric Series	$S = \dfrac{a_1}{1 - r}$ for $-1 < r < 1$. If $|r| \geq 1$, the infinite geometric series does not have a sum.

Example Find the sum of each infinite geometric series, if it exists.

a. $75 + 15 + 3 + \ldots$

First, find the value of r to determine if the sum exists. $a_1 = 75$ and $a_2 = 15$, so $r = \dfrac{15}{75}$ or $\dfrac{1}{5}$. Since $\left|\dfrac{1}{5}\right| < 1$, the sum exists. Now use the formula for the sum of an infinite geometric series.

$S = \dfrac{a_1}{1 - r}$ Sum formula

$= \dfrac{75}{1 - \dfrac{1}{5}}$ $a_1 = 75, r = \dfrac{1}{5}$

$= \dfrac{75}{\dfrac{4}{5}}$ or 93.75 Simplify.

The sum of the series is 93.75.

b. $\displaystyle\sum_{n=1}^{\infty} 48\left(-\dfrac{1}{3}\right)^{n-1}$

In this infinite geometric series, $a_1 = 48$ and $r = -\dfrac{1}{3}$.

$S = \dfrac{a_1}{1 - r}$ Sum formula

$= \dfrac{48}{1 - \left(-\dfrac{1}{3}\right)}$ $a_1 = 48, r = -\dfrac{1}{3}$

$= \dfrac{48}{\dfrac{4}{3}}$ or 36 Simplify.

Thus $\displaystyle\sum_{n=1}^{\infty} 48\left(-\dfrac{1}{3}\right)^{n-1} = 36$.

Exercises

Find the sum of each infinite geometric series, if it exists.

1. $a_1 = -7, r = \dfrac{5}{8}$

2. $1 + \dfrac{5}{4} + \dfrac{25}{16} + \ldots$

3. $a_1 = 4, r = \dfrac{1}{2}$

4. $\dfrac{2}{9} + \dfrac{5}{27} + \dfrac{25}{162} + \ldots$

5. $15 + 10 + 6\dfrac{2}{3} + \ldots$

6. $18 - 9 + 4\dfrac{1}{2} - 2\dfrac{1}{4} + \ldots$

7. $\dfrac{1}{10} + \dfrac{1}{20} + \dfrac{1}{40} + \ldots$

8. $1000 + 800 + 640 + \ldots$

9. $6 - 12 + 24 - 48 + \ldots$

10. $\displaystyle\sum_{n=1}^{\infty} 50\left(\dfrac{4}{5}\right)^{n-1}$

11. $\displaystyle\sum_{k=1}^{\infty} 22\left(-\dfrac{1}{2}\right)^{k-1}$

12. $\displaystyle\sum_{s=1}^{\infty} 24\left(\dfrac{7}{12}\right)^{s-1}$

11-5 Study Guide and Intervention (continued)

Infinite Geometric Series

Repeating Decimals A repeating decimal represents a fraction. To find the fraction, write the decimal as an infinite geometric series and use the formula for the sum.

Example Write each repeating decimal as a fraction.

a. $0.\overline{42}$

Write the repeating decimal as a sum.

$0.\overline{42} = 0.42424242...$

$= \dfrac{42}{100} + \dfrac{42}{10,000} + \dfrac{42}{1,000,000} + ...$

In this series $a_1 = \dfrac{42}{100}$ and $r = \dfrac{1}{100}$.

$S = \dfrac{a_1}{1 - r}$ Sum formula

$= \dfrac{\dfrac{42}{100}}{1 - \dfrac{1}{100}}$ $a_1 = \frac{42}{100}, r = \frac{1}{100}$

$= \dfrac{\dfrac{42}{100}}{\dfrac{99}{100}}$ Subtract.

$= \dfrac{42}{99}$ or $\dfrac{14}{33}$ Simplify.

Thus $0.\overline{42} = \dfrac{14}{33}$.

b. $0.5\overline{24}$

Let $S = 0.5\overline{24}$.

$S = 0.5242424...$	Write as a repeating decimal.
$1000S = 524.242424...$	Multiply each side by 1000.
$10S = 5.242424...$	Mulitply each side by 10.
$990S = 519$	Subtract the third equation from the second equation.
$S = \dfrac{519}{990}$ or $\dfrac{173}{330}$	Simplify.

Thus, $0.5\overline{24} = \dfrac{173}{330}$

Exercises

Write each repeating decimal as a fraction.

1. $0.\overline{2}$ 2. $0.\overline{8}$ 3. $0.\overline{30}$ 4. $0.\overline{87}$

5. $0.1\overline{0}$ 6. $0.\overline{54}$ 7. $0.\overline{75}$ 8. $0.1\overline{8}$

9. $0.\overline{62}$ 10. $0.7\overline{2}$ 11. $0.0\overline{72}$ 12. $0.0\overline{45}$

13. $0.0\overline{6}$ 14. $0.0\overline{138}$ 15. $0.0\overline{138}$ 16. $0.0\overline{81}$

17. $0.2\overline{45}$ 18. $0.4\overline{36}$ 19. $0.5\overline{4}$ 20. $0.8\overline{63}$

11-6 Study Guide and Intervention

Recursion and Special Sequences

Special Sequences In a **recursive formula**, each succeeding term is formulated from one or more previous terms. A recursive formula for a sequence has two parts:

1. the value(s) of the first term(s), and

2. an equation that shows how to find each term from the term(s) before it.

Example Find the first five terms of the sequence in which $a_1 = 6$, $a_2 = 10$, and $a_n = 2a_{n-2}$ for $n \geq 3$.

$a_1 = 6$

$a_2 = 10$

$a_3 = 2a_1 = 2(6) = 12$

$a_4 = 2a_2 = 2(10) = 20$

$a_5 = 2a_3 = 2(12) = 24$

The first five terms of the sequence are 6, 10, 12, 20, 24.

Exercises

Find the first five terms of each sequence.

1. $a_1 = 1$, $a_2 = 1$, $a_n = 2(a_{n-1} + a_{n-2})$, $n \geq 3$

2. $a_1 = 1$, $a_n = \dfrac{1}{1 + a_{n-1}}$, $n \geq 2$

3. $a_1 = 3$, $a_n = a_{n-1} + 2(n-2)$, $n \geq 2$

4. $a_1 = 5$, $a_n = a_{n-1} + 2$, $n \geq 2$

5. $a_1 = 1$, $a_n = (n-1)a_{n-1}$, $n \geq 2$

6. $a_1 = 7$, $a_n = 4a_{n-1} - 1$, $n \geq 2$

7. $a_1 = 3$, $a_2 = 4$, $a_n = 2a_{n-2} + 3a_{n-1}$, $n \geq 3$

8. $a_1 = 0.5$, $a_n = a_{n-1} + 2n$, $n \geq 2$

9. $a_1 = 8$, $a_2 = 10$, $a_n = \dfrac{a_{n-2}}{a_{n-1}}$, $n \geq 3$

10. $a_1 = 100$, $a_n = \dfrac{a_{n-1}}{n}$, $n \geq 2$

11-6 Study Guide and Intervention (continued)

Recursion and Special Sequences

Iteration Combining composition of functions with the concept of recursion leads to the process of **iteration**. Iteration is the process of composing a function with itself repeatedly.

Example Find the first three iterates of $f(x) = 4x - 5$ for an initial value of $x_0 = 2$.

To find the first iterate, find the value of the function for $x_0 = 2$

$$x_1 = f(x_0) \qquad \text{Iterate the function.}$$
$$ = f(2) \qquad x_0 = 2$$
$$ = 4(2) - 5 \text{ or } 3 \qquad \text{Simplify.}$$

To find the second iteration, find the value of the function for $x_1 = 3$.

$$x_2 = f(x_1) \qquad \text{Iterate the function.}$$
$$ = f(3) \qquad x_1 = 3$$
$$ = 4(3) - 5 \text{ or } 7 \qquad \text{Simplify.}$$

To find the third iteration, find the value of the function for $x_2 = 7$.

$$x_3 = f(x_2) \qquad \text{Iterate the function.}$$
$$ = f(7) \qquad x_2 = 7$$
$$ = 4(7) - 5 \text{ or } 23 \qquad \text{Simplify.}$$

The first three iterates are 3, 7, and 23.

Exercises

Find the first three iterates of each function for the given initial value.

1. $f(x) = x - 1; x_0 = 4$ **2.** $f(x) = x^2 - 3x; x_0 = 1$ **3.** $f(x) = x^2 + 2x + 1; x_0 = -2$

4. $f(x) = 4x - 6; x_0 = -5$ **5.** $f(x) = 6x - 2; x_0 = 3$ **6.** $f(x) = 100 - 4x; x_0 = -5$

7. $f(x) = 3x - 1; x_0 = 47$ **8.** $f(x) = x^3 - 5x^2; x_0 = 1$ **9.** $f(x) = 10x - 25; x_0 = 2$

10. $f(x) = 4x^2 - 9; x_0 = -1$ **11.** $f(x) = 2x^2 + 5; x_0 = -4$ **12.** $f(x) = \dfrac{x-1}{x+2}; x_0 = 1$

13. $f(x) = \frac{1}{2}(x + 11); x_0 = 3$ **14.** $f(x) = \dfrac{3}{x}; x_0 = 9$ **15.** $f(x) = x - 4x^2; x_0 = 1$

16. $f(x) = x + \dfrac{1}{x}; x_0 = 2$ **17.** $f(x) = x^3 - 5x^2 + 8x - 10;$ **18.** $f(x) = x^3 - x^2; x_0 = -2$
$ x_0 = 1$

11-7 Study Guide and Intervention

The Binomial Theorem

Pascal's Triangle Pascal's triangle is the pattern of coefficients of powers of binomials displayed in triangular form. Each row begins and ends with 1 and each coefficient is the sum of the two coefficients above it in the previous row.

Pascal's Triangle		
$(a + b)^0$	1
$(a + b)^1$	1 1
$(a + b)^2$	1 2 1
$(a + b)^3$	1 3 3 1
$(a + b)^4$	1 4 6 4 1
$(a + b)^5$	1 5 10 10 5 1

Example **Use Pascal's triangle to find the number of possible sequences consisting of 3 as and 2 bs.**

The coefficient 10 of the a^3b^2-term in the expansion of $(a + b)^5$ gives the number of sequences that result in three as and two bs.

Exercises

Expand each power using Pascal's triangle.

1. $(a + 5)^4$

2. $(x - 2y)^6$

3. $(j - 3k)^5$

4. $(2s + t)^7$

5. $(2p + 3q)^6$

6. $\left(a - \dfrac{b}{2}\right)^4$

7. Ray tosses a coin 15 times. How many different sequences of tosses could result in 4 heads and 11 tails?

8. There are 9 true/false questions on a quiz. If twice as many of the statements are true as false, how many different sequences of true/false answers are possible?

151

11-7 Study Guide and Intervention (continued)

The Binomial Theorem

The Binomial Theorem

Binomial Theorem	If n is a nonnegative integer, then $(a + b)^n = 1a^nb^0 + \dfrac{n}{1}a^{n-1}b^1 + \dfrac{n(n-1)}{1 \cdot 2}a^{n-2}b^2 + \dfrac{n(n-1)(n-2)}{1 \cdot 2 \cdot 3}a^{n-3}b^3 + \ldots + 1a^0b^n$

Another useful form of the Binomial Theorem uses **factorial** notation and sigma notation.

Factorial	If n is a positive integer, then $n! = n(n-1)(n-2) \cdot \ldots \cdot 2 \cdot 1$.
Binomial Theorem, Factorial Form	$(a + b)^n = \dfrac{n!}{n!0!}a^nb^0 + \dfrac{n!}{(n-1)!1!}a^{n-1}b^1 + \dfrac{n!}{(n-2)!2!}a^{n-2}b^2 + \ldots + \dfrac{n!}{0!n!}a^0b^n$ $= \displaystyle\sum_{k=0}^{n} \dfrac{n!}{(n-k)!k!}a^{n-k}b^k$

Example 1 Evaluate $\dfrac{11!}{8!}$.

$\dfrac{11!}{8!} = \dfrac{11 \cdot 10 \cdot 9 \cdot 8 \cdot 7 \cdot 6 \cdot 5 \cdot 4 \cdot 3 \cdot 2 \cdot 1}{8 \cdot 7 \cdot 6 \cdot 5 \cdot 4 \cdot 3 \cdot 2 \cdot 1}$

$= 11 \cdot 10 \cdot 9 = 990$

Example 2 Expand $(a - 3b)^4$.

$(a - 3b)^4 = \displaystyle\sum_{k=0}^{4} \dfrac{4!}{(4-k)!k!}a^{4-k}(-3b)^k$

$= \dfrac{4!}{4!0!}a^4 + \dfrac{4!}{3!1!}a^3(-3b)^1 + \dfrac{4!}{2!2!}a^2(-3b)^2 + \dfrac{4!}{1!3!}a(-3b)^3 + \dfrac{4!}{0!4!}(-3b)^4$

$= a^4 - 12a^3b + 54a^2b^2 - 108ab^3 + 81b^4$

Exercises

Evaluate each expression.

1. $5!$

2. $\dfrac{9!}{7!2!}$

3. $\dfrac{10!}{6!4!}$

Expand each power.

4. $(a - 3)^6$

5. $(r + 2s)^7$

6. $(4x + y)^4$

7. $\left(2 - \dfrac{m}{2}\right)^5$

Find the indicated term of each expansion.

8. third term of $(3x - y)^5$

9. fifth term of $(a + 1)^7$

10. fourth term of $(j + 2k)^8$

11. sixth term of $(10 - 3t)^7$

12. second term of $\left(m + \dfrac{2}{3}\right)^9$

13. seventh term of $(5x - 2)^{11}$

11-8 Study Guide and Intervention

Proof and Mathematical Induction

Mathematical Induction Mathematical induction is a method of proof used to prove statements about positive integers.

Mathematical Induction Proof	**Step 1** Show that the statement is true for some integer n. **Step 2** Assume that the statement is true for some positive integer k where $k \geq n$. This assumption is called the **inductive hypothesis**. **Step 3** Show that the statement is true for the next integer $k + 1$.

Example **Prove that $5 + 11 + 17 + \ldots + (6n - 1) = 3n^2 + 2n$.**

Step 1 When $n = 1$, the left side of the given equation is $6(1) - 1 = 5$. The right side is $3(1)^2 + 2(1) = 5$. Thus the equation is true for $n = 1$.

Step 2 Assume that $5 + 11 + 17 + \ldots + (6k - 1) = 3k^2 + 2k$ for some positive integer k.

Step 3 Show that the equation is true for $n = k + 1$. First, add $[6(k + 1) - 1]$ to each side.

$$5 + 11 + 17 + \ldots + (6k - 1) + [6(k + 1) - 1] = 3k^2 + 2k + [6(k + 1) - 1]$$

$$\begin{aligned}
&= 3k^2 + 2k + 6k + 5 && \text{Add.} \\
&= 3k^2 + 6k + 3 + 2k + 2 && \text{Rewrite.} \\
&= 3(k^2 + 2k + 1) + 2(k + 1) && \text{Factor.} \\
&= 3(k + 1)^2 + 2(k + 1) && \text{Factor.}
\end{aligned}$$

The last expression above is the right side of the equation to be proved, where n has been replaced by $k + 1$. Thus the equation is true for $n = k + 1$.

This proves that $5 + 11 + 17 + \ldots + (6n - 1) = 3n^2 + 2n$ for all positive integers n.

Exercises

Prove that each statement is true for all positive integers.

1. $3 + 7 + 11 + \ldots + (4n - 1) = 2n^2 + n$.

2. $500 + 100 + 20 + \ldots + 4 \cdot 5^{4 - n} = 625\left(1 - \dfrac{1}{5^n}\right)$.

11-8 Study Guide and Intervention (continued)

Proof and Mathematical Induction

Counterexamples To show that a formula or other generalization is *not* true, find a **counterexample**. Often this is done by substituting values for a variable.

Example 1 Find a counterexample for the formula $2n^2 + 2n + 3 = 2^{n+2} - 1$.

Check the first few positive integers.

n	Left Side of Formula	Right Side of Formula	
1	$2(1)^2 + 2(1) + 3 = 2 + 2 + 3$ or 7	$2^{1+2} - 1 = 2^3 - 1$ or 7	true
2	$2(2)^2 + 2(2) + 3 = 8 + 4 + 3$ or 15	$2^{2+2} - 1 = 2^4 - 1$ or 15	true
3	$2(3)^2 + 2(3) + 3 = 18 + 6 + 3$ or 27	$2^{3+2} - 1 = 2^5 - 1$ or 31	false

The value $n = 3$ provides a counterexample for the formula.

Example 2 Find a counterexample for the statement $x^2 + 4$ is either prime or divisible by 4.

n	$x^2 + 4$	True?	n	$x^2 + 4$	True?
1	1 + 4 or 5	Prime	6	36 + 4 or 40	Div. by 4
2	4 + 4 or 8	Div. by 4	7	49 + 4 or 53	Prime
3	9 + 4 or 13	Prime	8	64 + 4 or 68	Div. by 4
4	16 + 4 or 20	Div. by 4	9	81 + 4 or 85	Neither
5	25 + 4 or 29	Prime			

The value $n = 9$ provides a counterexample.

Exercises

Find a counterexample for each statement.

1. $1 + 5 + 9 + \ldots + (4n - 3) = 4n - 3$

2. $100 + 110 + 120 + \ldots + (10n + 90) = 5n^2 + 95$

3. $900 + 300 + 100 + \ldots + 100(3^{3-n}) = 900 \cdot \dfrac{2n}{n+1}$

4. $x^2 + x + 1$ is prime.

5. $2n + 1$ is a prime number.

6. $7n - 5$ is a prime number.

7. $\dfrac{1}{2} + 1 + \dfrac{3}{2} + \ldots + \dfrac{n}{2} = n - \dfrac{1}{2}$

8. $5n^2 + 1$ is divisible by 3.

9. $n^2 - 3n + 1$ is prime for $n > 2$.

10. $4n^2 - 1$ is divisible by either 3 or 5.

12-1 Study Guide and Intervention

The Counting Principle

Independent Events If the outcome of one event does not affect the outcome of another event and vice versa, the events are called **independent events**.

Fundamental Counting Principle	If event M can occur in m ways and is followed by event N that can occur in n ways, then the event M followed by the event N can occur in $m \cdot n$ ways.

Example **FOOD** For the Breakfast Special at the Country Pantry, customers can choose their eggs scrambled, fried, or poached, whole wheat or white toast, and either orange, apple, tomato, or grapefruit juice. How many different Breakfast Specials can a customer order?

A customer's choice of eggs does not affect his or her choice of toast or juice, so the events are independent. There are 3 ways to choose eggs, 2 ways to choose toast, and 4 ways to choose juice. By the Fundamental Counting Principle, there are $3 \cdot 2 \cdot 4$ or 24 ways to choose the Breakfast Special.

Exercises

Solve each problem.

1. The Palace of Pizza offers small, medium, or large pizzas with 14 different toppings available. How many different one-topping pizzas do they serve?

2. The letters A, B, C, and D are used to form four-letter passwords for entering a computer file. How many passwords are possible if letters can be repeated?

3. A restaurant serves 5 main dishes, 3 salads, and 4 desserts. How many different meals could be ordered if each has a main dish, a salad, and a dessert?

4. Marissa brought 8 T-shirts and 6 pairs of shorts to summer camp. How many different outfits consisting of a T-shirt and a pair of shorts does she have?

5. There are 6 different packages available for school pictures. The studio offers 5 different backgrounds and 2 different finishes. How many different options are available?

6. How many 5-digit even numbers can be formed using the digits 4, 6, 7, 2, 8 if digits can be repeated?

7. How many license plate numbers consisting of three letters followed by three numbers are possible when repetition is allowed?

8. How many 4-digit positive even integers are there?

12-1 Study Guide and Intervention (continued)
The Counting Principle

Dependent Events If the outcome of an event *does* affect the outcome of another event, the two events are said to be **dependent**. The Fundamental Counting Principle still applies.

Example   ENTERTAINMENT **The guests at a sleepover brought 8 videos. They decided they would only watch 3 videos. How many orders of 3 different videos are possible?**

After the group chooses to watch a video, they will not choose to watch it again, so the choices of videos are dependent events.

There are 8 choices for the first video. That leaves 7 choices for the second. After they choose the first 2 videos, there are 6 remaining choices. Thus by the Fundamental Counting Principle, there are $8 \cdot 7 \cdot 6$ or 336 orders of 3 different videos.

Exercises

Solve each problem.

1. Three students are scheduled to give oral reports on Monday. In how many ways can their presentations be ordered?

2. In how many ways can the first five letters of the alphabet be arranged if each letter is used only once?

3. In how many different ways can 4 different books be arranged on the shelf?

4. How many license plates consisting of three letters followed by three numbers are possible when no repetition is allowed?

5. Sixteen teams are competing in a soccer match. Gold, silver, and bronze medals will be awarded to the top three finishers. In how many ways can the medals be awarded?

6. In a word-building game each player picks 7 letter tiles. If Julio's letters are all different, how many 3-letter combinations can he make out of his 7 letters?

7. The editor has accepted 6 articles for the news letter. In how many ways can the 6 articles be ordered?

8. There are 10 one-hour workshops scheduled for the open house at the greenhouse. There is only one conference room available. In how many ways can the workshops be ordered?

9. The top 5 runners at the cross-country meet will receive trophies. If there are 22 runners in the race, in how many ways can the trophies be awarded?

12-2 Study Guide and Intervention

Permutations and Combinations

Permutations When a group of objects or people are arranged in a certain order, the arrangement is called a **permutation**.

Permutations	The number of permutations of n distinct objects taken r at a time is given by $P(n, r) = \dfrac{n!}{(n - r)!}$.
Permutations with Repetitions	The number of permutations of n objects of which p are alike and q are alike is $\dfrac{n!}{p!q!}$.

The rule for permutations with repetitions can be extended to any number of objects that are repeated.

Example From a list of 20 books, each student must choose 4 books for book reports. The first report is a traditional book report, the second a poster, the third a newspaper interview with one of the characters, and the fourth a timeline of the plot. How many different orderings of books can be chosen?

Since each book report has a different format, order is important. You must find the number of permutations of 20 objects taken 4 at a time.

$$P(n, r) = \frac{n!}{(n - r)!}$$
Permutation formula

$$P(20, 4) = \frac{20!}{(20 - 4)!}$$
$n = 20, r = 4$

$$= \frac{20!}{16!}$$
Simplify.

$$= \frac{20 \cdot 19 \cdot 18 \cdot 17 \cdot \overset{1}{\cancel{16}} \cdot \overset{1}{\cancel{15}} \cdot \ldots \cdot \overset{1}{\cancel{1}}}{\underset{1}{\cancel{16}} \cdot \underset{1}{\cancel{15}} \cdot \ldots \cdot \underset{1}{\cancel{1}}}$$
Divide by common factors.

$$= 116,280$$

Books for the book reports can be chosen 116,280 ways.

Exercises

Evaluate each expression.

1. $P(6, 3)$ **2.** $P(8, 5)$ **3.** $P(9, 4)$ **4.** $P(11, 6)$

How many different ways can the letters of each word be arranged?

5. MOM   **6.** MONDAY   **7.** STEREO

8. SCHOOL The high school chorus has been practicing 12 songs, but there is time for only 5 of them at the spring concert. How may different orderings of 5 songs are possible?

12-2 Study Guide and Intervention (continued)

Permutations and Combinations

Combinations An arrangement or selection of objects in which order is *not* important is called a combination.

Combinations	The number of combinations of n distinct objects taken r at a time is given by $C(n, r) = \dfrac{n!}{(n - r)!r!}$.

Example 1 **SCHOOL How many groups of 4 students can be selected from a class of 20?**

Since the order of choosing the students is not important, you must find the number of combinations of 20 students taken 4 at a time.

$C(n, r) = \dfrac{n!}{(n - r)!r!}$ Combination formula

$C(20, 4) = \dfrac{20!}{(20 - 4)!4!}$ $n = 20, r = 4$

$ = \dfrac{20!}{16!4!}$ or 4845

There are 4845 possible ways to choose 4 students.

Example 2 **In how many ways can you choose 1 vowel and 2 consonants from a set of 26 letter tiles? (Assume there are 5 vowels and 21 consonants.)**

By the Fundamental Counting Principle, you can multiply the number of ways to select one vowel and the number of ways to select 2 consonants. Only the letters chosen matter, not the order in which they were chosen, so use combinations.

$C(5, 1)$ One of 5 vowels are drawn.

$C(21, 2)$ Two of 21 consonants are drawn.

$C(5, 1) \cdot C(21, 2) = \dfrac{5!}{(5 - 1)!1!} \cdot \dfrac{21!}{(21 - 2)!2!}$ Combination formula

$ = \dfrac{5!}{4!} \cdot \dfrac{21!}{19!2!}$ Subtract.

$ = 5 \cdot 210$ or 1050   Simplify.

There are 1050 combinations of 1 vowel and 2 consonants.

Exercises

Evaluate each expression.

1. $C(5, 3)$ **2.** $C(7, 4)$ **3.** $C(15, 7)$ **4.** $C(10, 5)$

5. PLAYING CARDS From a standard deck of 52 cards, in how many ways can 5 cards be drawn?

6. HOCKEY How many hockey teams of 6 players can be formed from 14 players without regard to position played?

7. COMMITTEES From a group of 10 men and 12 women, how many committees of 5 men and 6 women can be formed?

12-3 Study Guide and Intervention

Probability

Probability and Odds In probability, a desired outcome is called a **success**; any other outcome is called a **failure**.

Probability of Success and Failure	If an event can succeed in s ways and fail in f ways, then the probabilities of success, $P(S)$, and of failure, $P(F)$, are as follows. $P(S) = \dfrac{s}{s+f}$ and $P(F) = \dfrac{f}{s+f}$.
Definition of Odds	If an event can succeed in s ways and fail in f ways, then the odds of success and of failure are as follows. Odds of success = $s:f$ Odds of failure = $f:s$

Example 1 **When 3 coins are tossed, what is the probability that at least 2 are heads?**

You can use a tree diagram to find the sample space.

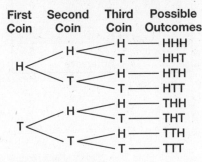

First Coin	Second Coin	Third Coin	Possible Outcomes

Of the 8 possible outcomes, 4 have at least 2 heads. So the probability of tossing at least 2 heads is $\dfrac{4}{8}$ or $\dfrac{1}{2}$.

Example 2 **What is the probability of picking 4 fiction books and 2 biographies from a best-seller list that consists of 12 fiction books and 6 biographies?**

By the Fundamental Counting Principle, the number of successes is $C(12, 4) \cdot C(6, 2)$. The total number of selections, $s + f$, of 6 books is $C(18, 6)$.

$$P(4 \text{ fiction}, 2 \text{ biography}) = \frac{C(12, 4) \cdot C(6, 2)}{C(18, 6)} \text{ or about } 0.40$$

The probability of selecting 4 fiction books and 2 biographies is about 40%.

Exercises

Find the odds of an event occurring, given the probability of the event.

1. $\dfrac{3}{7}$ 2. $\dfrac{4}{5}$ 3. $\dfrac{2}{13}$ 4. $\dfrac{1}{15}$

Find the probability of an event occurring, given the odds of the event.

5. 10:1   6. 2:5   7. 4:9   8. 8:3

One bag of candy contains 15 red candies, 10 yellow candies, and 6 green candies. Find the probability of each selection.

9. picking a red candy

10. not picking a yellow candy

11. picking a green candy

12. not picking a red candy

12-3 Study Guide and Intervention (continued)

Probability

Probability Distributions A **random variable** is a variable whose value is the numerical outcome of a random event. A **probability distribution** for a particular random variable is a function that maps the sample space to the probabilities of the outcomes in the sample space.

Example Suppose two dice are rolled. The table and the relative-frequency histogram show the distribution of the absolute value of the difference of the numbers rolled. Use the graph to determine which outcome is the most likely. What is its probability?

Difference	0	1	2	3	4	5
Probability	$\frac{1}{6}$	$\frac{5}{18}$	$\frac{2}{9}$	$\frac{1}{6}$	$\frac{1}{9}$	$\frac{1}{18}$

The greatest probability in the graph is $\frac{5}{18}$.

The most likely outcome is a difference of 1 and its probability is $\frac{5}{18}$.

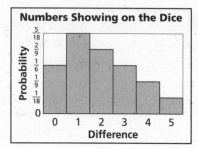

Numbers Showing on the Dice

Exercises

Four coins are tossed.

1. Complete the table below to show the probability distribution of the number of heads.

Number of Heads	0	1	2	3	4
Probability					

2. Make relative-frequency distribution of the data.

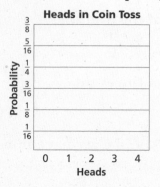

Heads in Coin Toss

NAME _____ DATE _____ PERIOD _____

12-4 Study Guide and Intervention
Multiplying Probabilities

Probability of Independent Events

Probability of Two Independent Events	If two events, A and B, are independent, then the probability of both occurring is $P(A \text{ and } B) = P(A) \cdot P(B)$.

Example In a board game each player has 3 different-colored markers. To move around the board the player first spins a spinner to determine which piece can be moved. He or she then rolls a die to determine how many spaces that colored piece should move. On a given turn what is the probability that a player will be able to move the yellow piece more than 2 spaces?

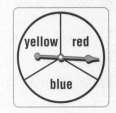

Let A be the event that the spinner lands on yellow, and let B be the event that the die shows a number greater than 2. The probability of A is $\frac{1}{3}$, and the probability of B is $\frac{2}{3}$.

$P(A \text{ and } B) = P(A) \cdot P(B)$ Probability of independent events

$= \frac{1}{3} \cdot \frac{2}{3}$ or $\frac{2}{9}$ Substitute and multiply.

The probability that the player can move the yellow piece more than 2 spaces is $\frac{2}{9}$.

Exercises

A die is rolled 3 times. Find the probability of each event.

1. a 1 is rolled, then a 2, then a 3

2. a 1 or a 2 is rolled, then a 3, then a 5 or a 6

3. 2 odd numbers are rolled, then a 6

4. a number less than 3 is rolled, then a 3, then a number greater than 3

5. A box contains 5 triangles, 6 circles, and 4 squares. If a figure is removed, replaced, and a second figure is picked, what is the probability that a triangle and then a circle will be picked?

6. A bag contains 5 red marbles and 4 white marbles. A marble is selected from the bag, then replaced, and a second selection is made. What is the probability of selecting 2 red marbles?

7. A jar contains 7 lemon jawbreakers, 3 cherry jawbreakers, and 8 rainbow jawbreakers. What is the probability of selecting 2 lemon jawbreakers in succession providing the jawbreaker drawn first is then replaced before the second is drawn?

Copyright © Glencoe/McGraw-Hill, a division of The McGraw-Hill Companies, Inc.

12-4 Study Guide and Intervention (continued)

Multiplying Probabilities

Probability of Dependent Events

Probability of Two Dependent Events	If two events, A and B, are dependent, then the probability of both events occurring is $P(A \text{ and } B) = P(A) \cdot P(B \text{ following } A)$.

Example 1 There are 7 dimes and 9 pennies in a wallet. Suppose two coins are to be selected at random, without replacing the first one. Find the probability of picking a penny and then a dime.

Because the coin is not replaced, the events are dependent.

Thus, $P(A \text{ and } B) = P(A) \cdot P(B \text{ following } A)$.

$P(\text{penny, then dime}) = P(\text{penny}) \cdot P(\text{dime following penny})$

$$\frac{9}{16} \cdot \frac{7}{15} = \frac{21}{80}$$

The probability is $\frac{21}{80}$ or about 0.26

Example 2 What is the probability of drawing, without replacement, 3 hearts, then a spade from a standard deck of cards?

Since the cards are not replaced, the events are dependent. Let H represent a heart and S represent a spade.

$P(\text{H, H, H, S}) = P(\text{H}) \cdot P(\text{H following H}) \cdot P(\text{H following 2 Hs}) \cdot P(\text{S following 3 Hs})$

$$= \frac{13}{52} \cdot \frac{12}{51} \cdot \frac{11}{50} \cdot \frac{13}{49} \text{ or about } 0.003$$

The probability is about 0.003 of drawing 3 hearts, then a spade.

Exercises

Find each probability.

1. The cup on Sophie's desk holds 4 red pens and 7 black pens. What is the probability of her selecting first a black pen, then a red one?

2. What is the probability of drawing two cards showing odd numbers from a set of cards that show the first 20 counting numbers if the first card is not replaced before the second is chosen?

3. There are 3 quarters, 4 dimes, and 7 nickels in a change purse. Suppose 3 coins are selected without replacement. What is the probability of selecting a quarter, then a dime, and then a nickel?

4. A basket contains 4 plums, 6 peaches, and 5 oranges. What is the probability of picking 2 oranges, then a peach if 3 pieces of fruit are selected at random?

5. A photographer has taken 8 black and white photographs and 10 color photographs for a brochure. If 4 photographs are selected at random, what is the probability of picking first 2 black and white photographs, then 2 color photographs?

12-5 Study Guide and Intervention

Adding Probabilities

Mutually Exclusive Events Events that cannot occur at the same time are called mutually exclusive events.

Probability of Mutually Exclusive Events	If two events, A and B, are mutually exclusive, then $P(A \text{ or } B) = P(A) + P(B)$.

This formula can be extended to any number of mutually exclusive events.

Example 1 To choose an afternoon activity, summer campers pull slips of paper out of a hat. Today there are 25 slips for a nature walk, 35 slips for swimming, and 30 slips for arts and crafts. What is the probability that a camper will pull a slip for a nature walk or for swimming?

These are mutually exclusive events. Note that there is a total of 90 slips.

$P(\text{nature walk or swimming}) = P(\text{nature walk}) + P(\text{swimming})$

$$= \frac{25}{90} + \frac{35}{90} \text{ or } \frac{2}{3}$$

The probability of a camper's pulling out a slip for a nature walk or for swimming is $\frac{2}{3}$.

Example 2 By the time one tent of 6 campers gets to the front of the line, there are only 10 nature walk slips and 15 swimming slips left. What is the probability that more than 4 of the 6 campers will choose a swimming slip?

$P(\text{more than 4 swimmers}) = P(5 \text{ swimmers}) + P(6 \text{ swimmers})$

$$= \frac{C(10, 1) \cdot C(15, 5)}{C(25, 6)} + \frac{C(10, 0) \cdot C(15, 6)}{C(25, 6)}$$

$$\approx 0.2$$

The probability of more than 4 of the campers swimming is about 0.2.

Exercises

Find each probability.

1. A bag contains 45 dyed eggs: 15 yellow, 12 green, and 18 red. What is the probability of selecting a green or a red egg?

2. The letters from the words LOVE and LIVE are placed on cards and put in a box. What is the probability of selecting an L or an O from the box?

3. A pair of dice is rolled, and the two numbers are added. What is the probability that the sum is either a 5 or a 7?

4. A bowl has 10 whole wheat crackers, 16 sesame crackers, and 14 rye crisps. If a person picks a cracker at random, what is the probability of picking either a sesame cracker or a rye crisp?

5. An art box contains 12 colored pencils and 20 pastels. If 5 drawing implements are chosen at random, what is the probability that at least 4 of them are pastels?

12-5 Study Guide and Intervention (continued)
Adding Probabilities

Inclusive Events

Probability of Inclusive Events	If two events, A and B, are inclusive, $P(A \text{ or } B) = P(A) + P(B) - P(A \text{ and } B)$.

Example What is the probability of drawing a face card or a black card from a standard deck of cards?

The two events are inclusive, since a card can be both a face card and a black card.

$P(\text{face card or black card}) = P(\text{face card}) + P(\text{black card}) - P(\text{black face card})$

$$= \frac{3}{13} + \frac{1}{2} - \frac{3}{26}$$

$$= \frac{8}{13} \text{ or about } 0.62$$

The probability of drawing either a face card or a black card is about 0.62

Exercises

Find each probability.

1. What is the probability of drawing a red card or an ace from a standard deck of cards?

2. Three cards are selected from a standard deck of 52 cards. What is the probability of selecting a king, a queen, or a red card?

3. The letters of the alphabet are placed in a bag. What is the probability of selecting a vowel or one of the letters from the word QUIZ?

4. A pair of dice is rolled. What is the probability that the sum is odd or a multiple of 3?

5. The Venn diagram at the right shows the number of juniors on varsity sports teams at Elmwood High School. Some athletes are on varsity teams for one season only, some athletes for two seasons, and some for all three seasons. If a varsity athlete is chosen at random from the junior class, what is the probability that he or she plays a fall or winter sport?

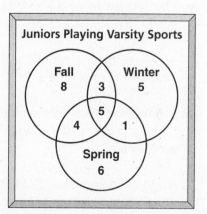

Juniors Playing Varsity Sports

Fall 8   3   Winter 5

5

4   1

Spring 6

12-6 Study Guide and Intervention

Statistical Measures

Measures of Central Tendency

Measures of Central Tendency	Use	When
	mean	the data are spread out and you want an average of values
	median	the data contain outliers
	mode	the data are tightly clustered around one or two values

Example Find the mean, median, and mode of the following set of data: {42, 39, 35, 40, 38, 35, 45}.

To find the mean, add the values and divide by the number of values.

$$\text{mean} = \frac{42 + 39 + 35 + 40 + 38 + 35 + 45}{7} \approx 39.14.$$

To find the median, arrange the values in ascending or descending order and choose the middle value. (If there is an even number of values, find the mean of the two middle values.) In this case, the median is 39.

To find the mode, take the most common value. In this case, the mode is 35.

Exercises

Find the mean, median, and mode of each set of data. Round to the nearest hundredth, if necessary.

1. {238, 261, 245, 249, 255, 262, 241, 245}

2. {9, 13, 8, 10, 11, 9, 12, 16, 10, 9}

3. {120, 108, 145, 129, 102, 132, 134, 118, 108, 142}

4. {68, 54, 73, 58, 63, 72, 65, 70, 61}

5. {34, 49, 42, 38, 40, 45, 34, 28, 43, 30}

6. The table at the right shows the populations of the six New England capitals. Which would be the most appropriate measure of central tendency to represent the data? Explain why and find that value.

Source: www.factfinder.census.gov

City	Population (rounded to the nearest 1000)
Augusta, ME	19,000
Boston, MA	589,000
Concord, NH	37,000
Hartford, CT	122,000
Montpelier, VT	8,000
Providence, RI	174,000

12-6 Study Guide and Intervention (continued)

Statistical Measures

Measures of Variation The *range* and the **standard deviation** measure how scattered a set of data is.

Standard Deviation	If a set of data consists of the n values x_1, x_2, ..., x_n and has mean $\bar{x}$, then the standard deviation is given by $\sigma = \sqrt{\dfrac{(x_1 - \bar{x})^2 + (x_2 - \bar{x})^2 + ... + (x_n - \bar{x})^2}{n}}$.

The square of the standard deviation is called the **variance**.

Example Find the variance and standard deviation of the data set {10, 9, 6, 9, 18, 4, 8, 20}.

Step 1 Find the mean.

$$\bar{x} = \frac{10 + 9 + 6 + 9 + 18 + 4 + 8 + 20}{8} = 10.5$$

Step 2 Find the variance.

$$\sigma^2 = \frac{(x_1 - \bar{x})^2 + (x_2 - \bar{x})^2 + ... + (x_n - \bar{x})^2}{n} \qquad \text{Standard variance formula}$$

$$= \frac{(10 - 10.5)^2 + (9 - 10.5)^2 + ... + (20 - 10.5)^2}{8}$$

$$= \frac{220}{8} \text{ or } 27.5$$

Step 3 Find the standard deviation.

$$\sigma = \sqrt{27.5}$$
$$\approx 5.2$$

The variance is 27.5 and the standard deviation is about 5.2.

Exercises

Find the variance and standard deviation of each set of data. Round to the nearest tenth.

1. {100, 89, 112, 104, 96, 108, 93}

2. {62, 54, 49, 62, 48, 53, 50}

3. {8, 9, 8, 8, 9, 7, 8, 9, 6}

4. {4.2, 5.0, 4.7, 4.5, 5.2, 4.8, 4.6, 5.1}

5. The table at the right lists the prices of ten brands of breakfast cereal. What is the standard deviation of the values to the nearest penny?

Price of 10 Brands of Breakfast Cereal	
$2.29	$3.19
$3.39	$2.79
$2.99	$3.09
$3.19	$2.59
$2.79	$3.29

12-7 Study Guide and Intervention

The Normal Distribution

Normal and Skewed Distributions A **continuous probability** distribution is represented by a curve.

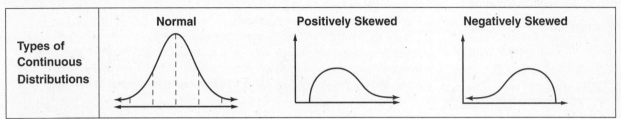

Types of Continuous Distributions	Normal	Positively Skewed	Negatively Skewed

Example Determine whether the data below appear to be *positively skewed*, *negatively skewed*, or *normally distributed*.

{100, 120, 110, 100, 110, 80, 100, 90, 100, 120, 100, 90, 110, 100, 90, 80, 100, 90}

Make a frequency table for the data.

Value	80	90	100	110	120
Frequency	2	4	7	3	2

Then use the data to make a histogram.

Since the histogram is roughly symmetric, the data appear to be normally distributed.

Exercises

Determine whether the data in each table appear to be *positively skewed*, *negatively skewed*, or *normally distributed*. Make a histogram of the data.

1. {27, 24, 29, 25, 27, 22, 24, 25, 29, 24, 25, 22, 27, 24, 22, 25, 24, 22}

2.

Shoe Size	4	5	6	7	8	9	10
No. of Students	1	2	4	8	5	1	2

3.

Housing Price	No. of Houses Sold
less than $100,000	0
$100,00–$120,000	1
$121,00–$140,000	3
$141,00–$160,000	7
$161,00–$180,000	8
$181,00–$200,000	6
over $200,000	12

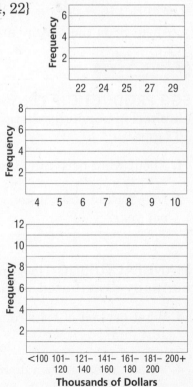

Thousands of Dollars

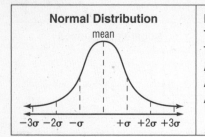

12-7 **Study Guide and Intervention** (continued)

The Normal Distribution

Use Normal Distributions

Normal Distribution	Normal distributions have these properties.
mean $-3\sigma\ -2\sigma\ -\sigma\quad +\sigma\ +2\sigma\ +3\sigma$	The graph is maximized at the mean. The mean, median, and mode are about equal. About 68% of the values are within one standard deviation of the mean. About 95% of the values are within two standard deviations of the mean. About 99% of the values are within three standard deviations of the mean.

Example   The heights of players in a basketball league are normally distributed with a mean of 6 feet 1 inch and a standard deviation of 2 inches.

a. What is the probability that a player selected at random will be shorter than 5 feet 9 inches?

Draw a normal curve. Label the mean and the mean plus or minus multiples of the standard deviation.

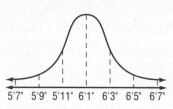

5'7" 5'9" 5'11" 6'1" 6'3" 6'5" 6'7"

The value of 5 feet 9 inches is 2 standard deviations below the mean, so approximately 2.5% of the players will be shorter than 5 feet 9 inches.

b. If there are 240 players in the league, about how many players are taller than 6 feet 3 inches?

The value of 6 feet 3 inches is one standard deviation above the mean. Approximately 16% of the players will be taller than this height.

$240 \times 0.16 \approx 38$

About 38 of the players are taller than 6 feet 3 inches.

Exercises

EGG PRODUCTION The number of eggs laid per year by a particular breed of chicken is normally distributed with a mean of 225 and a standard deviation of 10 eggs.

1. About what percent of the chickens will lay between 215 and 235 eggs per year?

2. In a flock of 400 chickens, about how many would you expect to lay more than 245 eggs per year?

MANUFACTURING The diameter of bolts produced by a manufacturing plant is normally distributed with a mean of 18 mm and a standard deviation of 0.2 mm.

3. What percent of bolts coming off of the assembly line have a diameter greater than 18.4 mm?

4. What percent have a diameter between 17.8 and 18.2 mm?

12-8 Study Guide and Intervention

Exponential and Binomial Distribution

Exponential Distribution

Exponential Distributions are used to predict the probabilities of events based on time.

Probability that a randomly chosen domain value for the exponential function $f(x)$ will be greater than the given value of x. The value m is the multiplicative inverse of the mean.	$f(x) = e^{-mx}$
Probability that a randomly chosen domain value for the exponential function $f(x)$ will be less than the given value of x. The value m is the multiplicative inverse of the mean.	$f(x) = 1 - e^{-mx}$

Example An exponential distribution function has a mean of 2. Graph the distribution function and label the mean. What is the probability that a randomly chosen value of x will be less than 3?

The equation for the function will be $f(x) = e^{-mx}$, where m is the multiplicative inverse of the mean. Since the mean is 2, the value of m will be $\frac{1}{2}$ or 0.5. Substituting the value of m into the equation for an exponential distribution, $f(x) = e^{-0.5x}$.

Since the function is applicable when x is greater than zero, the graph only includes the first quadrant x and y values. The y-axis represents the probability, which ranges from 0 to 1.

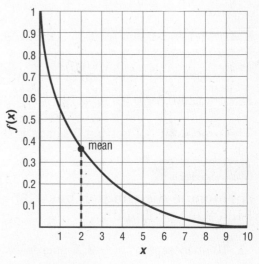

The probability that a randomly chosen value will be less than 3 can be found using the graph or the equation for the distribution. From the graph, the value of $f(x)$ is about 0.22 when x is 3, which means that the probability that x is greater than 3 is 0.22. Remember to subtract the value from one when you want to know the probability that a randomly chosen value will be less than the given value. The probability that x will be less than 3 is $1 - 0.22$, or 0.88.

To use the equation developed for the function to find the probability, substitute 3 for the value of x into the equation $f(x) = 1 - e^{-0.5x}$ and solve. The probability that x will be greater than 3 is 0.88.

Exercises

1. Write the equation for an exponential distribution that has a mean of 0.5.

An exponential distribution has a mean of 6. Find each probability.

2. $x > 7$ **3.** $x > 10$ **4.** $x > 2$ **5.** $x < 9$ **6.** $x < 7$

12-8 Study Guide and Intervention (continued)

Exponential and Binomial Distribution

Binomial Distribution

Binomial distributions occur when a series of trials only has two possible outcomes, success and failure, and the probability for success is the same in each trial.

Probability of x successes in n independent trials	$P(x) = C(n, x)\, p^x q^{n-x}$ where p is the probability of success of an individual trial and q is the probability of failure on that same individual trial ($p + q = 1$).
Expectation of a Binomial Distribution	$E(X) = np$, where n is the total number of trials and p is the probability of success.

Example You are going to flip a coin 10 times. What is the probability that 3 of the coin flips will result in heads?

A success in this case would be heads, and there is an equal chance that a flip will result in heads or tails, so the probability of success is 0.5.

$P(x) = C(n, x)\, p^x q^{n-x}$ Probability Formula

$P(x) = C(10, 3)\,(0.5)^3(0.5)^{10-3}$ $n = 10$, $x = 3$, $p = 0.5$

$P(x) = \dfrac{10!}{(10-3)!\,3!}(0.5)^3(0.5)^{10-3}$ Combination Formula

$P(x) = 120(0.5)^3(0.5)^7$ or 0.12 Simplify.

The probability of getting 3 heads is 0.12.

Exercises

A binomial distribution has a 30% rate of success. There are 20 trials. Find each probability.

1. 10 successes   2. 3 successes   3. 6 successes   4. 15 failures

The probability of getting a red fruit candy from an assorted bag is 0.1. There are 50 pieces of candy in a bag.

5. What is the probability that there are at least 4 pieces of red candy in the bag?

6. What is the probability that there are 8 pieces of red candy in the bag?

7. What is the probability that there are 2 or less pieces of red candy in the bag?

8. What is the expected number of red pieces of candy in the bag?

12-9 Study Guide and Intervention

Binomial Experiments

Binomial Expansions For situations with only 2 possible outcomes, you can use the Binomial Theorem to find probabilities. The coefficients of terms in a binomial expansion can be found by using combinations.

Example What is the probability that 3 coins show heads and 3 show tails when 6 coins are tossed?

There are 2 possible outcomes that are equally likely: heads (H) and tails (T). The tosses of 6 coins are independent events. When $(H + T)^6$ is expanded, the term containing H^3T^3, which represents 3 heads and 3 tails, is used to get the desired probability. By the Binomial Theorem the coefficient of H^3T^3 is $C(6, 3)$.

$$P(3 \text{ heads}, 3 \text{ tails}) = \frac{6!}{3!3!}\left(\frac{1}{2}\right)^3\left(\frac{1}{2}\right)^3 \quad P(H) = \tfrac{1}{2} \text{ and } P(T) = \tfrac{1}{2}$$

$$= \frac{20}{64}$$

$$= \frac{5}{16}$$

The probability of getting 3 heads and 3 tails is $\frac{5}{16}$ or 0.3125.

Exercises

Find each probability if a coin is tossed 8 times.

1. $P(\text{exactly 5 heads})$

2. $P(\text{exactly 2 heads})$

3. $P(\text{even number of heads})$

4. $P(\text{at least 6 heads})$

Mike guesses on all 10 questions of a true-false test. If the answers true and false are evenly distributed, find each probability.

5. Mike gets exactly 8 correct answers.

6. Mike gets at most 3 correct answers.

7. A die is tossed 4 times. What is the probability of tossing exactly two sixes?

12-9 Study Guide and Intervention (continued)

Binomial Experiments

Binomial Experiments

Binomial Experiments	A binomial experiment is possible if and only if all of these conditions occur. • There are exactly two outcomes for each trial. • There is a fixed number of trials. • The trials are independent. • The probabilities for each trial are the same.

Example Suppose a coin is weighted so that the probability of getting heads in any one toss is 90%. What is the probability of getting exactly 7 heads in 8 tosses?

The probability of getting heads is $\frac{9}{10}$, and the probability of getting tails is $\frac{1}{10}$. There are $C(8, 7)$ ways to choose the 7 heads.

$$P(7 \text{ heads}) = C(8, 7)\left(\frac{9}{10}\right)^7\left(\frac{1}{10}\right)^1$$

$$= 8 \cdot \frac{9^7}{10^8}$$

$$\approx 0.38$$

The probability of getting 7 heads in 8 tosses is about 38%.

Exercises

1. **BASKETBALL** For any one foul shot, Derek has a probability of 0.72 of getting the shot in the basket. As part of a practice drill, he shoots 8 shots from the foul line.

   a. What is the probability that he gets in exactly 6 foul shots?

   b. What is the probability that he gets in at least 6 foul shots?

2. **SCHOOL** A teacher is trying to decide whether to have 4 or 5 choices per question on her multiple choice test. She wants to prevent students who just guess from scoring well on the test.

   a. On a 5-question multiple-choice test with 4 choices per question, what is the probability that a student can score at least 60% by guessing?

   b. What is the probability that a student can score at least 60% by guessing on a test of the same length with 5 choices per question?

3. Julie rolls two dice and adds the two numbers.

   a. What is the probability that the sum will be divisible by 3?

   b. If she rolls the dice 5 times what is the chance that she will get exactly 3 sums that are divisible by 3?

4. **SKATING** During practice a skater falls 15% of the time when practicing a triple axel. During one practice session he attempts 20 triple axels.

   a. What is the probability that he will fall only once?

   b. What is the probability that he will fall 4 times?

12-10 Study Guide and Intervention

Sampling and Error

Bias A sample of size n is random (or **unbiased**) when every possible sample of size n has an equal chance of being selected. If a sample is biased, then information obtained from it may not be reliable.

Example **To find out how people in the U.S. feel about mass transit, people at a commuter train station are asked their opinion. Does this situation represent a random sample?**

No; the sample includes only people who actually use a mass-transit facility. The sample does not include people who ride bikes, drive cars, or walk.

Exercises

Determine whether each situation would produce a random sample. Write *yes* or *no* and explain your answer.

1. asking people in Phoenix, Arizona, about rainfall to determine the average rainfall for the United States

2. obtaining the names of tree types in North America by surveying all of the U.S. National Forests

3. surveying every tenth person who enters the mall to find out about music preferences in that part of the country

4. interviewing country club members to determine the average number of televisions per household in the community

5. surveying all students whose ID numbers end in 4 about their grades and career counseling needs

6. surveying parents at a day care facility about their preferences for brands of baby food for a marketing campaign

7. asking people in a library about the number of magazines to which they subscribe in order to describe the reading habits of a town

12-10 Study Guide and Intervention (continued)

Sampling and Error

Margin of Error The **margin of sampling error** gives a limit on the difference between how a sample responds and how the total population would respond.

Margin of Error	If the percent of people in a sample responding in a certain way is p and the size of the sample is n, then 95% of the time, the percent of the population responding in that same way will be between $p - ME$ and $p + ME$, where $ME = 2\sqrt{\dfrac{p(1-p)}{n}}$.

Example 1 In a survey of 4500 randomly selected voters, 62% favored candidate A. What is the margin of error?

$$ME = 2\sqrt{\frac{p(1-p)}{n}} \qquad \text{Formula for margin of sampling error}$$

$$= 2\sqrt{\frac{0.62 \cdot (1 - 0.62)}{4500}} \qquad p = 62\% \text{ or } 0.62, \, n = 4500$$

$$\approx 0.01447 \qquad \text{Use a calculator.}$$

The margin of error is about 1%. This means that there is a 95% chance that the percent of voters favoring candidate A is between $62 - 1$ or 61% and $62 + 1$ or 63%.

Example 2 The CD that 32% of teenagers surveyed plan to buy next is the latest from the popular new group BFA. If the margin of error of the survey is 2%, how many teenagers were surveyed?

$$ME = 2\sqrt{\frac{p(1-p)}{n}} \qquad \text{Formula for margin of sampling error}$$

$$0.02 = 2\sqrt{\frac{0.32 \cdot (1 - 0.32)}{n}} \qquad ME = 0.02, \, p = 0.32$$

$$0.01 = \sqrt{\frac{0.32(0.68)}{n}} \qquad \text{Divide each side by 2.}$$

$$0.0001 = \frac{0.32(0.68)}{n} \qquad \text{Square each side.}$$

$$n = \frac{0.32(0.68)}{0.0001} \qquad \text{Multiply by } n \text{ and divide by 0.0001}$$

$$n = 2176$$

2176 teenagers were surveyed.

Exercises

Find the margin of sampling error to the nearest percent.

1. $p = 45\%, n = 350$ **2.** $p = 12\%, n = 1500$ **3.** $p = 86\%, n = 600$

4. A study of 50,000 drivers in Indiana, Illinois, and Ohio showed that 68% preferred a speed limit of 75 mph over 65 mph on highways and country roads. What was the margin of sampling error to the nearest tenth of a percent?

13-1 Study Guide and Intervention

Right Triangle Trigonometry

Trigonometric Values

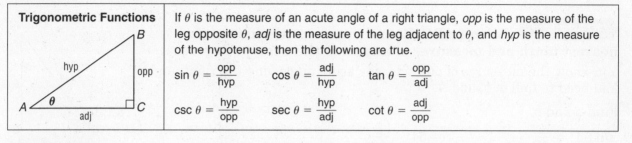

Trigonometric Functions	If θ is the measure of an acute angle of a right triangle, *opp* is the measure of the leg opposite θ, *adj* is the measure of the leg adjacent to θ, and *hyp* is the measure of the hypotenuse, then the following are true.
	$\sin \theta = \dfrac{opp}{hyp} \qquad \cos \theta = \dfrac{adj}{hyp} \qquad \tan \theta = \dfrac{opp}{adj}$
	$\csc \theta = \dfrac{hyp}{opp} \qquad \sec \theta = \dfrac{hyp}{adj} \qquad \cot \theta = \dfrac{adj}{opp}$

Example **Find the values of the six trigonometric functions for angle θ.**

Use the Pythagorean Theorem to find x, the measure of the leg opposite θ.

$x^2 + 7^2 = 9^2$ Pythagorean Theorem
$x^2 + 49 = 81$ Simplify.
$\quad x^2 = 32$ Subtract 49 from each side.
$\quad\quad x = \sqrt{32} \text{ or } 4\sqrt{2}$ Take the square root of each side.

Use opp = $4\sqrt{2}$, adj = 7, and hyp = 9 to write each trigonometric ratio.

$\sin \theta = \dfrac{4\sqrt{2}}{9} \qquad \cos \theta = \dfrac{7}{9} \qquad \tan \theta = \dfrac{4\sqrt{2}}{7} \qquad \csc \theta = \dfrac{9\sqrt{2}}{8} \qquad \sec \theta = \dfrac{9}{7} \qquad \cot \theta = \dfrac{7\sqrt{2}}{8}$

Exercises

Find the values of the six trigonometric functions for angle θ.

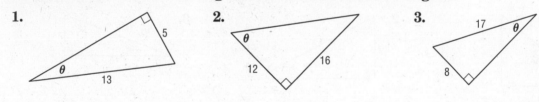

1.

2.

3.

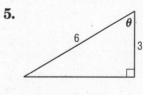

4.

5.

6.

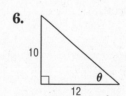

13-1 Study Guide and Intervention (continued)

Right Triangle Trigonometry

Right Triangle Problems

Example Solve $\triangle ABC$. Round measures of sides to the nearest tenth and measures of angles to the nearest degree.

You know the measures of one side, one acute angle, and the right angle. You need to find a, b, and A.

Find a and b.

$$\sin 54° = \frac{b}{18} \qquad \cos 54° = \frac{a}{18}$$
$$b = 18 \sin 54° \qquad a = 18 \cos 54°$$
$$b \approx 14.6 \qquad a \approx 10.6$$

Find A.

$$54° + A = 90° \qquad \text{Angles } A \text{ and } B \text{ are complementary.}$$
$$A = 36° \qquad \text{Solve for } A.$$

Therefore $A = 36°$, $a \approx 10.6$, and $b \approx 14.6$.

Exercises

Write an equation involving sin, cos, or tan that can be used to find x. Then solve the equation. Round measures of sides to the nearest tenth.

1.

2.

3.

Solve $\triangle ABC$ by using the given measurements. Round measures of sides to the nearest tenth and measures of angles to the nearest degree.

4. $A = 80°$, $b = 6$

5. $B = 25°$, $c = 20$

6. $b = 8$, $c = 14$

7. $a = 6$, $b = 7$

8. $a = 12$, $B = 42°$

9. $a = 15$, $A = 54°$

NAME _____ DATE _____ PERIOD _____

13-2 Study Guide and Intervention
Angles and Angle Measurement

Angle Measurement An angle is determined by two rays. The degree measure of an angle is described by the amount and direction of rotation from the **initial side** along the positive x-axis to the **terminal side**. A counterclockwise rotation is associated with positive angle measure and a clockwise rotation is associated with negative angle measure. An angle can also be measured in **radians**.

Radian and Degree Measure	To rewrite the radian measure of an angle in degrees, multiply the number of radians by $\frac{180°}{\pi \text{ radians}}$.
	To rewrite the degree measure of an angle in radians, multiply the number of degrees by $\frac{\pi \text{ radians}}{180°}$.

Example 1 Draw an angle with measure 290° in standard notation.

The negative y-axis represents a positive rotation of 270°. To generate an angle of 290°, rotate the terminal side 20° more in the counterclockwise direction.

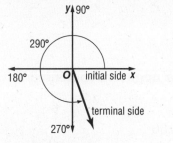

Example 2 Rewrite the degree measure in radians and the radian measure in degrees.

a. 45°

$$45° = 45°\left(\frac{\pi \text{ radians}}{180°}\right) = \frac{\pi}{4} \text{ radians}$$

b. $\frac{5\pi}{3}$ radians

$$\frac{5\pi}{3} \text{ radians} = \frac{5\pi}{3}\left(\frac{180°}{\pi}\right) = 300°$$

Exercises

Draw an angle with the given measure in standard position.

1. 160° 2. $-\frac{5\pi}{4}$ 3. 400°

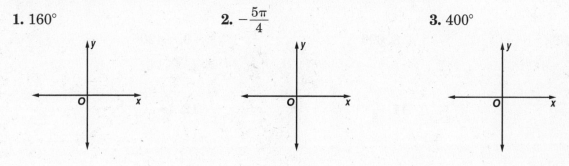

Rewrite each degree measure in radians and each radian measure in degrees.

4. 140° 5. −860° 6. $-\frac{3\pi}{5}$ 7. $\frac{11\pi}{3}$

13-2 Study Guide and Intervention (continued)

Angles and Angle Measurement

Coterminal Angles When two angles in standard position have the same terminal sides, they are called **coterminal angles**. You can find an angle that is coterminal to a given angle by adding or subtracting a multiple of 360°. In radian measure, a coterminal angle is found by adding or subtracting a multiple of 2π.

Example Find one angle with positive measure and one angle with negative measure coterminal with each angle.

a. 250°

A positive angle is $250° + 360°$ or $610°$.

A negative angle is $250° - 360°$ or $-110°$.

b. $\dfrac{5\pi}{8}$

A positive angle is $\dfrac{5\pi}{8} + 2\pi$ or $\dfrac{21\pi}{8}$.

A negative angle is $\dfrac{5\pi}{8} - 2\pi$ or $-\dfrac{11\pi}{8}$.

Exercises

Find one angle with a positive measure and one angle with a negative measure coterminal with each angle.

1. $65°$

2. $-75°$

3. $230°$

4. $420°$

5. $340°$

6. $-130°$

7. $-290°$

8. $690°$

9. $-420°$

10. $\dfrac{\pi}{9}$

11. $\dfrac{3\pi}{8}$

12. $\dfrac{6\pi}{5}$

13. $\dfrac{-7\pi}{4}$

14. $\dfrac{15\pi}{4}$

15. $\dfrac{-13\pi}{6}$

16. $\dfrac{17\pi}{5}$

17. $\dfrac{-5\pi}{3}$

18. $\dfrac{-11\pi}{4}$

13-3 Study Guide and Intervention

Trigonometric Functions of General Angles

Trigonometric Functions and General Angles

Trigonometric Functions, θ in Standard Position	Let θ be an angle in standard position and let $P(x, y)$ be a point on the terminal side of θ. By the Pythagorean Theorem, the distance r from the origin is given by $r = \sqrt{x^2 + y^2}$. The trigonometric functions of an angle in standard position may be defined as follows.
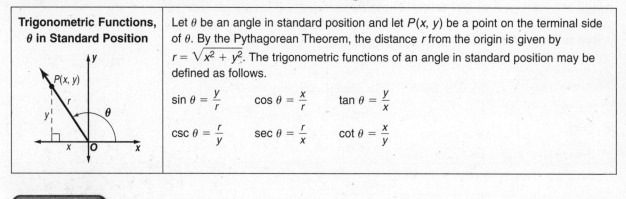	$\sin \theta = \dfrac{y}{r}$ $\cos \theta = \dfrac{x}{r}$ $\tan \theta = \dfrac{y}{x}$ $\csc \theta = \dfrac{r}{y}$ $\sec \theta = \dfrac{r}{x}$ $\cot \theta = \dfrac{x}{y}$

Example Find the exact values of the six trigonometric functions of θ if the terminal side of θ contains the point $(-5, 5\sqrt{2})$.

You know that $x = -5$ and $y = 5\sqrt{2}$. You need to find r.

$r = \sqrt{x^2 + y^2}$ Pythagorean Theorem

$\quad = \sqrt{(-5)^2 + (5\sqrt{2})^2}$ Replace x with -5 and y with $5\sqrt{2}$.

$\quad = \sqrt{75}$ or $5\sqrt{3}$

Now use $x = -5$, $y = 5\sqrt{2}$, and $r = 5\sqrt{3}$ to write the ratios.

$\sin \theta = \dfrac{y}{r} = \dfrac{5\sqrt{2}}{5\sqrt{3}} = \dfrac{\sqrt{6}}{3}$ $\cos \theta = \dfrac{x}{r} = \dfrac{-5}{5\sqrt{3}} = -\dfrac{\sqrt{3}}{3}$ $\tan \theta = \dfrac{y}{x} = \dfrac{5\sqrt{2}}{-5} = -\sqrt{2}$

$\csc \theta = \dfrac{r}{y} = \dfrac{5\sqrt{3}}{5\sqrt{2}} = \dfrac{\sqrt{6}}{2}$ $\sec \theta = \dfrac{r}{x} = \dfrac{5\sqrt{3}}{-5} = -\sqrt{3}$ $\cot \theta = \dfrac{x}{y} = \dfrac{-5}{5\sqrt{2}} = -\dfrac{\sqrt{2}}{2}$

Exercises

Find the exact values of the six trigonometric functions of θ if the terminal side of θ in standard position contains the given point.

1. $(8, 4)$

2. $(4, 4\sqrt{3})$

3. $(0, -4)$

4. $(6, 2)$

13-3 **Study Guide and Intervention** *(continued)*

Trigonometric Functions of General Angles

Reference Angles If θ is a nonquadrantal angle in standard position, its reference angle θ' is defined as the acute angle formed by the terminal side of θ and the x-axis.

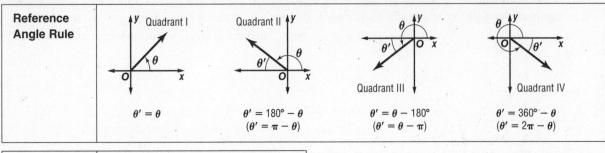

Reference Angle Rule				
	Quadrant I	Quadrant II	Quadrant III	Quadrant IV
	$\theta' = \theta$	$\theta' = 180° - \theta$ $(\theta' = \pi - \theta)$	$\theta' = \theta - 180°$ $(\theta' = \theta - \pi)$	$\theta' = 360° - \theta$ $(\theta' = 2\pi - \theta)$

		Quadrant			
Signs of Trigonometric Functions	**Function**	**I**	**II**	**III**	**IV**
	$\sin \theta$ or $\csc \theta$	+	+	−	−
	$\cos\theta$ or $\sec \theta$	+	−	−	+
	$\tan \theta$ or $\cot \theta$	+	−	+	−

Example 1 **Sketch an angle of measure 205°. Then find its reference angle.**

Because the terminal side of 205° lies in Quadrant III, the reference angle θ' is 205° − 180° or 25°.

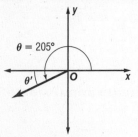

Example 2 **Use a reference angle to find the exact value of $\cos \dfrac{3\pi}{4}$.**

Because the terminal side of $\dfrac{3\pi}{4}$ lies in Quadrant II, the reference angle θ' is $\pi - \dfrac{3\pi}{4}$ or $\dfrac{\pi}{4}$.

The cosine function is negative in Quadrant II.

$$\cos \frac{3\pi}{4} = -\cos \frac{\pi}{4} = -\frac{\sqrt{2}}{2}.$$

Exercises

Find the exact value of each trigonometric function.

1. $\tan(-510°)$

2. $\csc \dfrac{11\pi}{4}$

3. $\sin(-90°)$

4. $\cot 1665°$

5. $\cot 30°$

6. $\tan 315°$

7. $\csc \dfrac{\pi}{4}$

8. $\tan \dfrac{4\pi}{3}$

13-4 Study Guide and Intervention

Law of Sines

Law of Sines The area of any triangle is one half the product of the lengths of two sides and the sine of the included angle.

Area of a Triangle	area $= \frac{1}{2} bc \sin A$ area $= \frac{1}{2} ac \sin B$ area $= \frac{1}{2} ab \sin C$	

You can use the Law of Sines to solve any triangle if you know the measures of two angles and any side, or the measures of two sides and the angle opposite one of them.

Law of Sines	$\dfrac{\sin A}{a} = \dfrac{\sin B}{b} = \dfrac{\sin C}{c}$

Example 1 Find the area of $\triangle ABC$ if $a = 10$, $b = 14$, and $C = 40°$.

Area $= \frac{1}{2}ab \sin C$ Area formula

     $= \frac{1}{2}(10)(14)\sin 40°$ Replace a, b, and C.

     ≈ 44.9951 Use a calculator.

The area of the triangle is approximately 45 square units.

Example 2 If $a = 12$, $b = 9$, and $A = 28°$, find B.

$\dfrac{\sin A}{a} = \dfrac{\sin B}{b}$ Law of Sines

$\dfrac{\sin 28°}{12} = \dfrac{\sin B}{9}$ Replace A, a, and b.

$\sin B = \dfrac{9 \sin 28°}{12}$ Solve for $\sin B$.

$\sin B \approx 0.3521$ Use a calculator.

$B \approx 20.62°$ Use the $\sin^{-1}$ function.

Exercises

Find the area of $\triangle ABC$ to the nearest tenth.

1.
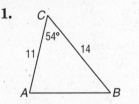

2.
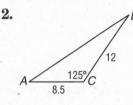

3.
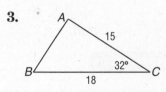

Solve each triangle. Round measures of sides to the nearest tenth and measures of angles to the nearest degree.

4. $B = 42°$, $C = 68°$, $a = 10$ **5.** $A = 40°$, $B = 14°$, $a = 52$ **6.** $A = 15°$, $B = 50°$, $b = 36$

13-4 Study Guide and Intervention (continued)

Law of Sines

One, Two, or No Solutions

Possible Triangles Given Two Sides and One Opposite Angle	Suppose you are given a, b, and A for a triangle.
	If a is acute:
	$a < b \sin A$ $\Rightarrow$ no solution
	$a = b \sin A$ $\Rightarrow$ one solution
	$b > a > b \sin A$ $\Rightarrow$ two solutions
	$a > b$ $\Rightarrow$ one solution
	If A is right or obtuse:
	$a \le b \Rightarrow$ no solution
	$a > b \Rightarrow$ one solution

Example Determine whether $\triangle ABC$ has no solutions, one solution, or two solutions. Then solve $\triangle ABC$.

a. $A = 48°$, $a = 11$, and $b = 16$

Since A is acute, find $b \sin A$ and compare it with a.
$b \sin A = 16 \sin 48° \approx 11.89$
Since $11 < 11.89$, there is no solution.

b. $A = 34°$, $a = 6$, $b = 8$

Since A is acute, find $b \sin A$ and compare it with a; $b \sin A = 8 \sin 34° \approx 4.47$. Since $8 > 6 > 4.47$, there are two solutions. Thus there are two possible triangles to solve.

Acute B

First use the Law of Sines to find B.

$$\frac{\sin B}{8} = \frac{\sin 34°}{6}$$

$$\sin B \approx 0.7456$$

$$B \approx 48°$$

The measure of angle C is about $180° - (34° + 48°)$ or about $98°$.

Use the Law of Sines again to find c.

$$\frac{\sin 98°}{c} \approx \frac{\sin 34°}{6}$$

$$c \approx \frac{6 \sin 98°}{\sin 34°}$$

$$c \approx 10.6$$

Obtuse B

To find B you need to find an obtuse angle whose sine is also 0.7456.

To do this, subtract the angle given by your calculator, $48°$, from $180°$. So B is approximately $132°$.

The measure of angle C is about $180° - (34° + 132°)$ or about $14°$.

Use the Law of Sines to find c.

$$\frac{\sin 14°}{c} \approx \frac{\sin 34°}{6}$$

$$c \approx \frac{6 \sin 14°}{\sin 34°}$$

$$c \approx 2.6$$

Exercises

Determine whether each triangle has no solutions, one solution, or two solutions. Then solve each triangle. Round measures of sides to the nearest tenth and measures of angles to the nearest degree.

1. $A = 50°$, $a = 34$, $b = 40$ **2.** $A = 24°$, $a = 3$, $b = 8$ **3.** $A = 125°$, $a = 22$, $b = 15$

13-5 Study Guide and Intervention
Law of Cosines

Law of Cosines

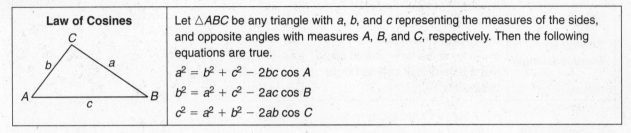

Law of Cosines	Let $\triangle ABC$ be any triangle with a, b, and c representing the measures of the sides, and opposite angles with measures A, B, and C, respectively. Then the following equations are true. $a^2 = b^2 + c^2 - 2bc \cos A$ $b^2 = a^2 + c^2 - 2ac \cos B$ $c^2 = a^2 + b^2 - 2ab \cos C$

You can use the Law of Cosines to solve any triangle if you know the measures of two sides and the included angle, or the measures of three sides.

Example Solve $\triangle ABC$.

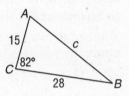

You are given the measures of two sides and the included angle.
Begin by using the Law of Cosines to determine c.
$c^2 = a^2 + b^2 - 2ab \cos C$
$c^2 = 28^2 + 15^2 - 2(28)(15)\cos 82°$
$c^2 \approx 892.09$
$c \approx 29.9$

Next you can use the Law of Sines to find the measure of angle A.
$\dfrac{\sin A}{a} = \dfrac{\sin C}{c}$
$\dfrac{\sin A}{28} \approx \dfrac{\sin 82°}{29.9}$
$\sin A \approx 0.9273$
$A \approx 68°$

The measure of B is about $180° - (82° + 68°)$ or about $30°$.

Exercises

Solve each triangle described below. Round measures of sides to the nearest tenth and angles to the nearest degree.

1. $a = 14, c = 20, B = 38°$

2. $A = 60°, c = 17, b = 12$

3. $a = 4, b = 6, c = 3$

4. $A = 103°, b = 31, c = 52$

5. $a = 15, b = 26, C = 132°$

6. $a = 31, b = 52, c = 43$

13-5 Study Guide and Intervention *(continued)*

Law of Cosines

Choose the Method

	Given	Begin by Using
Solving an Oblique Triangle	two angles and any side	Law of Sines
	two sides and a non-included angle	Law of Sines
	two sides and their included angle	Law of Cosines
	three sides	Law of Cosines

Example Determine whether $\triangle ABC$ should be solved by beginning with the Law of Sines or Law of Cosines. Then solve the triangle. Round the measure of the side to the nearest tenth and measures of angles to the nearest degree.

You are given the measures of two sides and their included angle, so use the Law of Cosines.

$a^2 = b^2 + c^2 - 2bc \cos A$ Law of Cosines

$a^2 = 20^2 + 8^2 - 2(20)(8) \cos 34°$ $b = 20, c = 8, A = 34°$

$a^2 \approx 198.71$ Use a calculator.

$a \approx 14.1$ Use a calculator.

Use the Law of Sines to find B.

$\dfrac{\sin B}{b} = \dfrac{\sin A}{a}$ Law of Sines

$\sin B \approx \dfrac{20 \sin 34°}{14.1}$ $b = 20, A = 34°, a \approx 14.1$

$B \approx 128°$ Use the $\sin^{-1}$ function.

The measure of angle C is approximately $180° - (34° + 128°)$ or about $18°$.

Exercises

Determine whether each triangle should be solved by beginning with the Law of Sines or Law of Cosines. Then solve each triangle. Round measures of sides to the nearest tenth and measures of angles to the nearest degree.

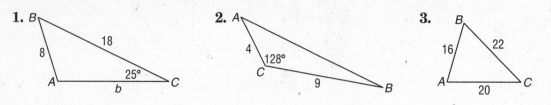

1. *B*, 8, 18, 25°, *A*, *b*, *C*

2. *A*, 4, 128°, *C*, 9, *B*

3. *B*, 16, 22, *A*, 20, *C*

4. $A = 58°, a = 12, b = 8.5$ **5.** $a = 28, b = 35, c = 20$ **6.** $A = 82°, B = 44°, b = 11$

13-6 Study Guide and Intervention

Circular Functions

Unit Circle Definitions

Definition of Sine and Cosine	If the terminal side of an angle θ in standard position intersects the unit circle at $P(x, y)$, then $\cos \theta = x$ and $\sin \theta = y$. Therefore, the coordinates of P can be written as $P(\cos \theta, \sin \theta)$. 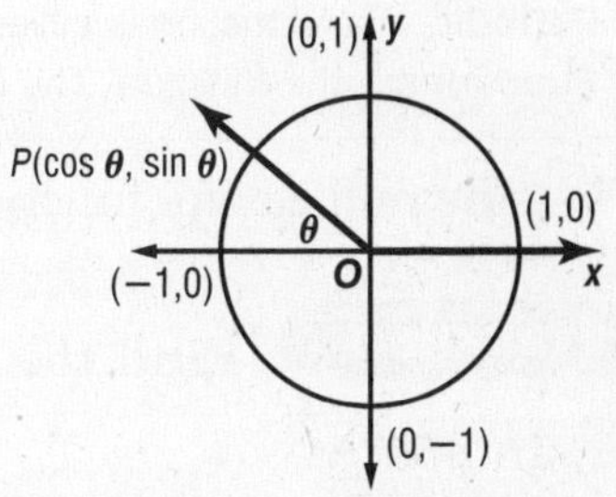	

Example Given an angle θ in standard position, if $P\left(-\dfrac{5}{6}, \dfrac{\sqrt{11}}{6}\right)$ lies on the terminal side and on the unit circle, find $\sin \theta$ and $\cos \theta$.

$P\left(-\dfrac{5}{6}, \dfrac{\sqrt{11}}{6}\right) = P(\cos \theta, \sin \theta)$, so $\sin \theta = \dfrac{\sqrt{11}}{6}$ and $\cos \theta = -\dfrac{5}{6}$.

Exercises

If θ is an angle in standard position and if the given point P is located on the terminal side of θ and on the unit circle, find $\sin \theta$ and $\cos \theta$.

1. $P\left(-\dfrac{\sqrt{3}}{2}, \dfrac{1}{2}\right)$

2. $P(0, -1)$

3. $P\left(-\dfrac{2}{3}, \dfrac{\sqrt{5}}{3}\right)$

4. $P\left(-\dfrac{4}{5}, -\dfrac{3}{5}\right)$

5. $P\left(\dfrac{1}{6}, -\dfrac{\sqrt{35}}{6}\right)$

6. $P\left(\dfrac{\sqrt{7}}{4}, \dfrac{3}{4}\right)$

7. P is on the terminal side of $\theta = 45°$.

8. P is on the terminal side of $\theta = 120°$.

9. P is on the terminal side of $\theta = 240°$.

10. P is on the terminal side of $\theta = 330°$.

13-6 Study Guide and Intervention *(continued)*
Circular Functions

Periodic Functions

Periodic Functions	A function is called **periodic** if there is a number a such that $f(x) = f(x + a)$ for all x in the domain of the function. The least positive value of a for which $f(x) = f(x + a)$ is called the period of the function.

The sine and cosine functions are periodic; each has a period of $360°$ or 2π.

Example 1 Find the exact value of each function.

a. sin 855°

$$\sin 855° = \sin(135° + 720°) = \sin 135° = \frac{\sqrt{2}}{2}$$

b. $\cos\left(\dfrac{31\pi}{6}\right)$

$$\cos\left(\frac{31\pi}{6}\right) = \cos\left(\frac{7\pi}{6} + 4\pi\right)$$

$$= \cos\frac{7\pi}{6} = -\frac{\sqrt{3}}{2}$$

Example 2 Determine the period of the function graphed below.

The pattern of the function repeats every 10 units, so the period of the function is 10.

Exercises

Find the exact value of each function.

1. $\cos(-240°)$ **2.** $\cos 2880°$ **3.** $\sin(-510°)$

4. $\sin 495°$ **5.** $\cos\left(-\dfrac{5\pi}{2}\right)$ **6.** $\sin\left(\dfrac{5\pi}{3}\right)$

7. $\cos\left(\dfrac{11\pi}{4}\right)$ **8.** $\sin\left(-\dfrac{3\pi}{4}\right)$ **9.** $\cos 1440°$

10. $\sin(-750°)$ **11.** $\cos 870°$ **12.** $\cos 1980°$

13. $\sin 7\pi$ **14.** $\sin\left(-\dfrac{13\pi}{4}\right)$ **15.** $\cos\left(\dfrac{23\pi}{6}\right)$

16. Determine the period of the function.

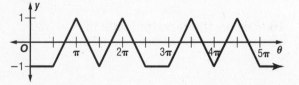

13-7 Study Guide and Intervention

Inverse Trigonometric Functions

Trigonometric Values You can use a calculator to find the values of trigonometric expressions.

Example Find each value. Write angle measures in radians. Round to the nearest hundredth.

a. Find $\tan \left(\text{Sin}^{-1} \frac{1}{2} \right)$.

Let $\theta = \text{Sin}^{-1} \frac{1}{2}$. Then $\text{Sin } \theta = \frac{1}{2}$ with $-\frac{\pi}{2} < \theta < \frac{\pi}{2}$. The value $\theta = \frac{\pi}{6}$ satisfies both

conditions. $\tan \frac{\pi}{6} = \frac{\sqrt{3}}{3}$ so $\tan \left(\text{Sin}^{-1} \frac{1}{2} \right) = \frac{\sqrt{3}}{3}$.

b. Find $\cos (\text{Tan}^{-1} 4.2)$.

KEYSTROKES: [COS] [2nd] [tan^{-1}] 4.2 [ENTER] .2316205273

Therefore $\cos (\text{Tan}^{-1} 4.2) \approx 0.23$.

Exercises

Find each value. Write angle measures in radians. Round to the nearest hundredth.

1. $\cot (\text{Tan}^{-1} 2)$

2. $\text{Arctan}(-1)$

3. $\cot^{-1} 1$

4. $\cos \left[\text{Sin}^{-1} \left(-\frac{\sqrt{2}}{2} \right) \right]$

5. $\text{Sin}^{-1} \left(-\frac{\sqrt{3}}{2} \right)$

6. $\sin \left(\text{Arcsin} \frac{\sqrt{3}}{2} \right)$

7. $\tan \left[\text{Arcsin} \left(-\frac{5}{7} \right) \right]$

8. $\sin \left(\text{Tan}^{-1} \frac{5}{12} \right)$

9. $\sin [\text{Arctan}^{-1} (-\sqrt{2})]$

10. $\text{Arccos} \left(-\frac{\sqrt{3}}{2} \right)$

11. $\text{Arcsin} \left(\frac{\sqrt{3}}{2} \right)$

12. $\text{Arccot} \left(-\frac{\sqrt{3}}{3} \right)$

13. $\cos [\text{Arcsin} (-0.7)]$

14. $\tan (\text{Cos}^{-1} 0.28)$

15. $\cos (\text{Arctan} 5)$

16. $\text{Sin}^{-1} (-0.78)$

17. $\text{Cos}^{-1} 0.42$

18. $\text{Arctan} (-0.42)$

19. $\sin (\text{Cos}^{-1} 0.32)$

20. $\cos (\text{Arctan} 8)$

21. $\tan (\text{Cos}^{-1} 0.95)$

13-7 Study Guide and Intervention (continued)

Inverse Trigonometric Functions

Solve Equations Using Inverses If the domains of trigonometric functions are restricted to their **principal values**, then their inverses are also functions.

Principal Values of Sine, Cosine, and Tangent	$y = \text{Sin } x$ if and only if $y = \sin x$ and $-\frac{\pi}{2} \le x \le \frac{\pi}{2}$. $y = \text{Cos } x$ if and only if $y = \cos x$ and $0 \le x \le \pi$. $y = \text{Tan } x$ if and only if $y = \tan x$ and $-\frac{\pi}{2} \le x \le \frac{\pi}{2}$.
Inverse Sine, Cosine, and Tangent	Given $y = \text{Sin } x$, the inverse Sine function is defined by $y = \text{Sin}^{-1} x$ or $y = \text{Arcsin } x$. Given $y = \text{Cos } x$, the inverse Cosine function is defined by $y = \text{Cos}^{-1} x$ or $y = \text{Arccos } x$. Given $y = \text{Tan } x$, the inverse Tangent function is given by $y = \text{Tan}^{-1} x$ or $y = \text{Arctan } x$.

Example 1 Solve $x = \text{Sin}^{-1}\left(\frac{\sqrt{3}}{2}\right)$.

If $x = \text{Sin}^{-1}\left(\frac{\sqrt{3}}{2}\right)$, then $\text{Sin } x = \frac{\sqrt{3}}{2}$ and $-\frac{\pi}{2} \le x \le \frac{\pi}{2}$.

The only x that satisfies both criteria is $x = \frac{\pi}{3}$ or $60°$.

Example 2 Solve $\text{Arctan}\left(-\frac{\sqrt{3}}{3}\right) = x$.

If $x = \text{Arctan}\left(-\frac{\sqrt{3}}{3}\right)$, then $\text{Tan } x = -\frac{\sqrt{3}}{3}$ and $-\frac{\pi}{2} \le x \le \frac{\pi}{2}$.

The only x that satisfies both criteria is $-\frac{\pi}{6}$ or $-30°$.

Exercises

Solve each equation by finding the value of x to the nearest degree.

1. $\text{Cos}^{-1}\left(-\frac{\sqrt{3}}{2}\right) = x$

2. $x = \text{Sin}^{-1}\frac{\sqrt{3}}{2}$

3. $x = \text{Arccos}\,(-0.8)$

4. $x = \text{Arctan}\,\sqrt{3}$

5. $x = \text{Arccos}\left(-\frac{\sqrt{2}}{2}\right)$

6. $x = \text{Tan}^{-1}\,(-1)$

7. $\text{Sin}^{-1}\,0.45 = x$

8. $x = \text{Arcsin}\left(-\frac{\sqrt{3}}{2}\right)$

9. $x = \text{Arccos}\left(-\frac{1}{2}\right)$

10. $\text{Cos}^{-1}\,(-0.2) = x$

11. $x = \text{Tan}^{-1}\,(-\sqrt{3})$

12. $x = \text{Arcsin}\,0.3$

13. $x = \text{Tan}^{-1}\,(15)$

14. $x = \text{Cos}^{-1}\,1$

15. $\text{Arctan}^{-1}\,(-3) = x$

16. $x = \text{Sin}^{-1}\,(-0.9)$

17. $\text{Arccos}^{-1}\,0.15$

18. $x = \text{Tan}^{-1}\,0.2$

14-1 Study Guide and Intervention

Graphing Trigonometric Functions

Graph Trigonometric Functions To graph a trigonometric function, make a table of values for known degree measures (0°, 30°, 45°, 60°, 90°, and so on). Round function values to the nearest tenth, and plot the points. Then connect the points with a smooth, continuous curve. The *period* of the sine, cosine, secant, and cosecant functions is 360° or 2π radians.

Amplitude of a Function	The **amplitude** of the graph of a periodic function is the absolute value of half the difference between its maximum and minimum values.

Example Graph $y = \sin \theta$ for $-360° \le \theta \le 0°$.

First make a table of values.

θ	$-360°$	$-330°$	$-315°$	$-300°$	$-270°$	$-240°$	$-225°$	$-210°$	$-180°$
$\sin \theta$	0	$\dfrac{1}{2}$	$\dfrac{\sqrt{2}}{2}$	$\dfrac{\sqrt{3}}{2}$	1	$\dfrac{\sqrt{3}}{2}$	$\dfrac{\sqrt{2}}{2}$	$\dfrac{1}{2}$	0

θ	$-150°$	$-135°$	$-120°$	$-90°$	$-60°$	$-45°$	$-30°$	$0°$	
$\sin \theta$	$-\dfrac{1}{2}$	$-\dfrac{\sqrt{2}}{2}$	$-\dfrac{\sqrt{3}}{2}$	-1	$-\dfrac{\sqrt{3}}{2}$	$-\dfrac{\sqrt{2}}{2}$	$-\dfrac{1}{2}$	0	

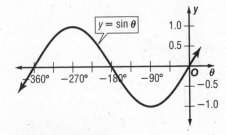

Exercises

Graph the following functions for the given domain.

1. $\cos \theta$, $-360° \le \theta \le 0°$

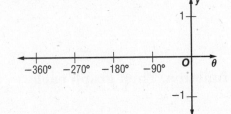

2. $\tan \theta$, $-2\pi \le \theta \le 0$

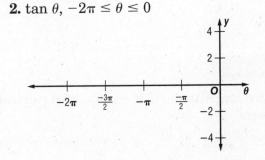

What is the amplitude of each function?

3.

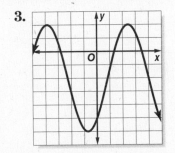

4.

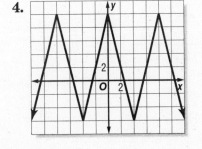

189

14-1 Study Guide and Intervention (continued)

Graphing Trigonometric Functions

Variations of Trigonometric Functions

Amplitudes and Periods	For functions of the form $y = a \sin b\theta$ and $y = a \cos b\theta$, the amplitude is $|a|$, and the period is $\frac{360°}{|b|}$ or $\frac{2\pi}{|b|}$.
	For functions of the form $y = a \tan b\theta$, the amplitude is not defined, and the period is $\frac{180°}{|b|}$ or $\frac{\pi}{|b|}$.

Example Find the amplitude and period of each function. Then graph the function.

a. $y = 4 \cos \dfrac{\theta}{3}$

First, find the amplitude.

$|a| = |4|$, so the amplitude is 4.

Next find the period.

$\dfrac{360°}{\left|\frac{1}{3}\right|} = 1080°$

Use the amplitude and period to help graph the function.

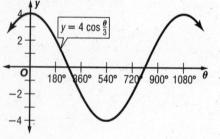

b. $y = -\dfrac{1}{2} \tan 2\theta$

The amplitude is not defined, and the period is $\dfrac{\pi}{2}$.

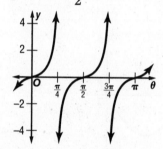

Exercises

Find the amplitude, if it exists, and period of each function. Then graph each function.

1. $y = -3 \sin \theta$

2. $y = 2 \tan \dfrac{\theta}{2}$

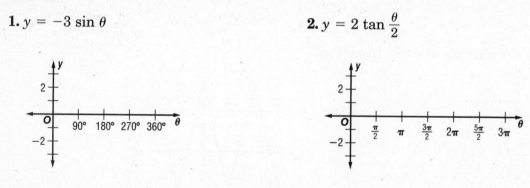

14-2 Study Guide and Intervention

Translations of Trigonometric Graphs

Horizontal Translations When a constant is subtracted from the angle measure in a trigonometric function, a **phase shift** of the graph results.

Phase Shift	The horizontal phase shift of the graphs of the functions $y = a \sin b(\theta - h)$, $y = a \cos b(\theta - h)$, and $y = a \tan b(\theta - h)$ is h, where $b > 0$. If $h > 0$, the shift is to the right. If $h < 0$, the shift is to the left.

Example State the amplitude, period, and phase shift for $y = \dfrac{1}{2} \cos 3\left(\theta - \dfrac{\pi}{2}\right)$. Then graph the function.

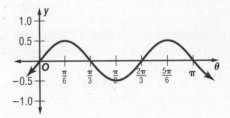

Amplitude: $a = \left|\dfrac{1}{2}\right|$ or $\dfrac{1}{2}$

Period: $\dfrac{2\pi}{|b|} = \dfrac{2\pi}{|3|}$ or $\dfrac{2\pi}{3}$

Phase Shift: $h = \dfrac{\pi}{2}$

The phase shift is to the right since $\dfrac{\pi}{2} > 0$.

Exercises

State the amplitude, period, and phase shift for each function. Then graph the function.

1. $y = 2 \sin (\theta + 60°)$

2. $y = \tan \left(\theta - \dfrac{\pi}{2}\right)$

3. $y = 3 \cos (\theta - 45°)$

4. $y = \dfrac{1}{2} \sin 3\left(\theta - \dfrac{\pi}{3}\right)$

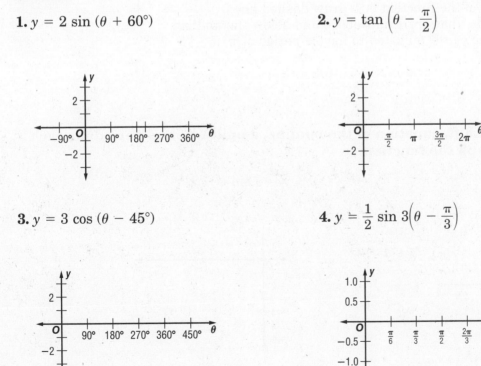

14-2 Study Guide and Intervention (continued)

Translations of Trigonometric Graphs

Vertical Translations When a constant is added to a trigonometric function, the graph is shifted vertically.

Vertical Shift	The vertical shift of the graphs of the functions $y = a \sin b(\theta - h) + k$, $y = a \cos b(\theta - h) + k$, and $y = a \tan b(\theta - h) + k$ is k. If $k > 0$, the shift is up. If $k < 0$, the shift is down.

The **midline** of a vertical shift is $y = k$.

Graphing Trigonometric Functions	Step 1	Determine the vertical shift, and graph the midline.
	Step 2	Determine the amplitude, if it exists. Use dashed lines to indicate the maximum and minimum values of the function.
	Step 3	Determine the period of the function and graph the appropriate function.
	Step 4	Determine the phase shift and translate the graph accordingly.

Example State the vertical shift, equation of the midline, amplitude, and period for $y = \cos 2\theta - 3$. Then graph the function.

Vertical Shift: $k = -3$, so the vertical shift is 3 units down.

The equation of the midline is $y = -3$.

Amplitude: $|a| = |1|$ or 1

Period: $\dfrac{2\pi}{|b|} = \dfrac{2\pi}{|2|}$ or π

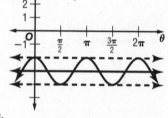

Since the amplitude of the function is 1, draw dashed lines parallel to the midline that are 1 unit above and below the midline. Then draw the cosine curve, adjusted to have a period of π.

Exercises

State the vertical shift, equation of the midline, amplitude, and period for each function. Then graph the function.

1. $y = \dfrac{1}{2} \cos \theta + 2$

2. $y = 3 \sin \theta - 2$

192

14-3 Study Guide and Intervention

Trigonometric Identities

Find Trigonometric Values A **trigonometric identity** is an equation involving trigonometric functions that is true for all values for which every expression in the equation is defined.

Basic Trigonometric Identities	Quotient Identities	$\tan \theta = \dfrac{\sin \theta}{\cos \theta}$	$\cot \theta = \dfrac{\cos \theta}{\sin \theta}$	
	Reciprocal Identities	$\csc \theta = \dfrac{1}{\sin \theta}$	$\sec \theta = \dfrac{1}{\cos \theta}$	$\cot \theta = \dfrac{1}{\tan \theta}$
	Pythagorean Identities	$\cos^2 \theta + \sin^2 \theta = 1$	$\tan^2 \theta + 1 = \sec^2 \theta$	$\cot^2 \theta + 1 = \csc^2 \theta$

Example Find the value of $\cot \theta$ if $\csc \theta = -\dfrac{11}{5}$; $180° < \theta < 270°$.

$\cot^2 \theta + 1 = \csc^2 \theta$ Trigonometric identity

$\cot^2 \theta + 1 = \left(-\dfrac{11}{5}\right)^2$ Substitute $-\dfrac{11}{5}$ for $\csc \theta$.

$\cot^2 \theta + 1 = \dfrac{121}{25}$ Square $-\dfrac{11}{5}$.

$\cot^2 \theta = \dfrac{96}{25}$ Subtract 1 from each side.

$\cot \theta = \pm\dfrac{4\sqrt{6}}{5}$ Take the square root of each side.

Since θ is in the third quadrant, $\cot \theta$ is positive, Thus $\cot \theta = \dfrac{4\sqrt{6}}{5}$.

Exercises

Find the value of each expression.

1. $\tan \theta$, if $\cot \theta = 4$; $180° < \theta < 270°$

2. $\csc \theta$, if $\cos \theta = \dfrac{\sqrt{3}}{2}$; $0° \leq \theta < 90°$

3. $\cos \theta$, if $\sin \theta = \dfrac{3}{5}$; $0° \leq \theta < 90°$

4. $\sec \theta$, if $\sin \theta = \dfrac{1}{3}$; $0° \leq \theta < 90°$

5. $\cos \theta$, if $\tan \theta = -\dfrac{4}{3}$; $90° < \theta < 180°$

6. $\tan \theta$, if $\sin \theta = \dfrac{3}{7}$; $0° \leq \theta < 90°$

7. $\sec \theta$, if $\cos \theta = -\dfrac{7}{8}$; $90° < \theta < 180°$

8. $\sin \theta$, if $\cos \theta = \dfrac{6}{7}$; $270° \leq \theta < 360°$

9. $\cot \theta$, if $\csc \theta = \dfrac{12}{5}$; $90° < \theta < 180°$

10. $\sin \theta$, if $\csc \theta = -\dfrac{9}{4}$; $270° < \theta < 360°$

14-3 Study Guide and Intervention *(continued)*

Trigonometric Identities

Simplify Expressions The simplified form of a trigonometric expression is written as a numerical value or in terms of a single trigonometric function, if possible. Any of the trigonometric identities on page 849 can be used to simplify expressions containing trigonometric functions.

Example 1 Simplify $(1 - \cos^2 \theta) \sec \theta \cot \theta + \tan \theta \sec \theta \cos^2 \theta$.

$$(1 - \cos^2 \theta) \sec \theta \cot \theta + \tan \theta \sec \theta \cos^2 \theta = \sin^2 \theta \cdot \frac{1}{\cos \theta} \cdot \frac{\cos \theta}{\sin \theta} + \frac{\sin \theta}{\cos \theta} \cdot \frac{1}{\cos \theta} \cdot \cos^2 \theta$$

$$= \sin \theta + \sin \theta$$

$$= 2 \sin \theta$$

Example 2 Simplify $\dfrac{\sec \theta \cdot \cot \theta}{1 - \sin \theta} - \dfrac{\csc \theta}{1 + \sin \theta}$.

$$\frac{\sec \theta \cdot \cot \theta}{1 - \sin \theta} - \frac{\csc \theta}{1 + \sin \theta} = \frac{\dfrac{1}{\cos \theta} \cdot \dfrac{\cos \theta}{\sin \theta}}{1 - \sin \theta} - \frac{\dfrac{1}{\sin \theta}}{1 + \sin \theta}$$

$$= \frac{\dfrac{1}{\sin \theta}(1 + \sin \theta) - \dfrac{1}{\sin \theta}(1 - \sin \theta)}{(1 - \sin \theta)(1 + \sin \theta)}$$

$$= \frac{\dfrac{1}{\sin \theta} + 1 - \dfrac{1}{\sin \theta} + 1}{1 - \sin^2 \theta}$$

$$= \frac{2}{\cos^2 \theta}$$

Exercises

Simplify each expression.

1. $\dfrac{\tan \theta \cdot \csc \theta}{\sec \theta}$

2. $\dfrac{\sin \theta \cdot \cot \theta}{\sec^2 \theta - \tan^2 \theta}$

3. $\dfrac{\sin^2 \theta - \cot \theta \cdot \tan \theta}{\cot \theta \cdot \sin \theta}$

4. $\dfrac{\cos \theta}{\sec \theta - \tan \theta}$

5. $\dfrac{\tan \theta \cdot \cos \theta}{\sin \theta} + \cot \theta \cdot \sin \theta \cdot \tan \theta \cdot \csc \theta$

6. $\dfrac{\csc^2 \theta - \cot^2 \theta}{\tan \theta \cdot \cos \theta}$

7. $3 \tan \theta \cdot \cot \theta + 4 \sin \theta \cdot \csc \theta + 2 \cos \theta \cdot \sec \theta$

8. $\dfrac{1 - \cos^2 \theta}{\tan \theta \cdot \sin \theta}$

14-4 Study Guide and Intervention

Verifying Trigonometric Identities

Transform One Side of an Equation Use the basic trigonometric identities along with the definitions of the trigonometric functions to verify trigonometric identities. Often it is easier to begin with the more complicated side of the equation and transform that expression into the form of the simpler side.

Example Verify that each of the following is an identity.

a. $\dfrac{\sin \theta}{\cot \theta} - \sec \theta = -\cos \theta$

Transform the left side.

$\dfrac{\sin \theta}{\cot \theta} - \sec \theta \stackrel{?}{=} -\cos \theta$

$\dfrac{\sin \theta}{\frac{\cos \theta}{\sin \theta}} - \dfrac{1}{\cos \theta} \stackrel{?}{=} -\cos \theta$

$\dfrac{\sin^2 \theta}{\cos \theta} - \dfrac{1}{\cos \theta} \stackrel{?}{=} -\cos \theta$

$\dfrac{\sin^2 - 1}{\cos \theta} \stackrel{?}{=} -\cos \theta$

$\dfrac{-\cos^2 \theta}{\cos \theta} \stackrel{?}{=} -\cos \theta$

$-\cos \theta = -\cos \theta$

b. $\dfrac{\tan \theta}{\csc \theta} + \cos \theta = \sec \theta$

Transform the left side.

$\dfrac{\tan \theta}{\csc \theta} + \cos \theta \stackrel{?}{=} \sec \theta$

$\dfrac{\frac{\sin \theta}{\cos \theta}}{\frac{1}{\sin \theta}} + \cos \theta \stackrel{?}{=} \sec \theta$

$\dfrac{\sin^2 \theta}{\cos \theta} + \cos \theta \stackrel{?}{=} \sec \theta$

$\dfrac{\sin^2 \theta + \cos^2 \theta}{\cos \theta} \stackrel{?}{=} \sec \theta$

$\dfrac{1}{\cos \theta} \stackrel{?}{=} \sec \theta$

$\sec \theta = \sec \theta$

Exercises

Verify that each of the following is an identity.

1. $1 + \csc^2 \theta \cdot \cos^2 \theta = \csc^2 \theta$

2. $\dfrac{\sin \theta}{1 - \cos \theta} - \dfrac{\cot \theta}{1 + \cos \theta} = \dfrac{1 - \cos^3 \theta}{\sin^3 \theta}$

14-4 Study Guide and Intervention (continued)

Verifying Trigonometric Identities

Transform Both Sides of an Equation The following techniques can be helpful in verifying trigonometric identities.

- Substitute one or more basic identities to simplify an expression.
- Factor or multiply to simplify an expression.
- Multiply both numerator and denominator by the same trigonometric expression.
- Write each side of the identity in terms of sine and cosine only. Then simplify each side.

Example Verify that $\dfrac{\tan^2 \theta + 1}{\sin \theta \cdot \tan \theta \cdot \sec \theta + 1} = \sec^2 \theta - \tan^2 \theta$ is an identity.

$$\dfrac{\tan^2 \theta + 1}{\sin \theta \cdot \tan \theta \cdot \sec \theta + 1} \overset{?}{=} \sec^2 \theta - \tan^2 \theta$$

$$\dfrac{\sec^2 \theta}{\sin \theta \cdot \dfrac{\sin \theta}{\cos \theta} \cdot \dfrac{1}{\cos \theta} + 1} \overset{?}{=} \dfrac{1}{\cos^2 \theta} - \dfrac{\sin^2 \theta}{\cos^2 \theta}$$

$$\dfrac{\dfrac{1}{\cos^2 \theta}}{\dfrac{\sin^2 \theta}{\cos^2 \theta} + 1} \overset{?}{=} \dfrac{1 - \sin^2 \theta}{\cos^2 \theta}$$

$$\dfrac{\dfrac{1}{\cos^2 \theta}}{\dfrac{\sin^2 \theta + \cos^2 \theta}{\cos^2 \theta}} \overset{?}{=} \dfrac{\cos^2 \theta}{\cos^2 \theta}$$

$$\dfrac{1}{\sin^2 \theta + \cos^2 \theta} \overset{?}{=} 1$$

$$1 = 1$$

Exercises

Verify that each of the following is an identity.

1. $\csc \theta \cdot \sec \theta = \cot \theta + \tan \theta$

2. $\dfrac{\tan^2 \theta}{1 - \cos^2 \theta} = \dfrac{\sec \theta}{\cos \theta}$

3. $\dfrac{\cos \theta \cdot \cot \theta}{\sin \theta} = \dfrac{\csc \theta}{\sin \theta \cdot \sec^2 \theta}$

4. $\dfrac{\csc^2 \theta - \cot^2 \theta}{\sec^2 \theta} = \cot^2 \theta(1 - \cos^2 \theta)$

14-5 Study Guide and Intervention

Sum and Difference of Angles Formulas

Sum and Difference Formulas The following formulas are useful for evaluating an expression like sin 15° from the known values of sine and cosine of 60° and 45°.

Sum and Difference of Angles	The following identities hold true for all values of α and β. $\cos(\alpha \pm \beta) = \cos\alpha \cdot \cos\beta \mp \sin\alpha \cdot \sin\beta$ $\sin(\alpha \pm \beta) = \sin\alpha \cdot \cos\beta \pm \cos\alpha \cdot \sin\beta$

Example Find the exact value of each expression.

a. cos 345°

$$\cos 345° = \cos(300° + 45°)$$
$$= \cos 300° \cdot \cos 45° - \sin 300° \cdot \sin 45°$$
$$= \frac{1}{2} \cdot \frac{\sqrt{2}}{2} - \left(-\frac{\sqrt{3}}{2}\right) \cdot \frac{\sqrt{2}}{2}$$
$$= \frac{\sqrt{2} + \sqrt{6}}{4}$$

b. sin (−105°)

$$\sin(-105°) = \sin(45° - 150°)$$
$$= \sin 45° \cdot \cos 150° - \cos 45° \cdot \sin 150°$$
$$= \frac{\sqrt{2}}{2} \cdot \left(-\frac{\sqrt{3}}{2}\right) - \frac{\sqrt{2}}{2} \cdot \frac{1}{2}$$
$$= -\frac{\sqrt{2} + \sqrt{6}}{4}$$

Exercises

Find the exact value of each expression.

1. sin 105°   **2.** cos 285°   **3.** cos (−75°)

4. cos (−165°)   **5.** sin 195°   **6.** cos 420°

7. sin (−75°)   **8.** cos 135°   **9.** cos (−15°)

10. sin 345°   **11.** cos (−105°)   **12.** sin 495°

14-5 **Study Guide and Intervention** *(continued)*

Sum and Difference of Angles Formulas

Verify Identities You can also use the sum and difference of angles formulas to verify identities.

Example 1 Verify that $\cos\left(\theta + \dfrac{3\pi}{2}\right) = \sin\theta$ is an identity.

$$\cos\left(\theta + \dfrac{3\pi}{2}\right) \overset{?}{=} \sin\theta \qquad \text{Original equation}$$

$$\cos\theta \cdot \cos\dfrac{3\pi}{2} - \sin\theta \cdot \sin\dfrac{3\pi}{2} \overset{?}{=} \sin\theta \qquad \text{Sum of Angles Formula}$$

$$\cos\theta \cdot 0 - \sin\theta \cdot (-1) \overset{?}{=} \sin\theta \qquad \text{Evaluate each expression.}$$

$$\sin\theta = \sin\theta \qquad \text{Simplify.}$$

Example 2 Verify that $\sin\left(\theta - \dfrac{\pi}{2}\right) + \cos(\theta + \pi) = -2\cos\theta$ is an identity.

$$\sin\left(\theta - \dfrac{\pi}{2}\right) + \cos(\theta + \pi) \overset{?}{=} -2\cos\theta \qquad \text{Original equation}$$

$$\sin\theta \cdot \cos\dfrac{\pi}{2} - \cos\theta \cdot \sin\dfrac{\pi}{2} + \cos\theta \cdot \cos\pi - \sin\theta \cdot \sin\pi \overset{?}{=} -2\cos\theta \qquad \begin{array}{l}\text{Sum and Difference of}\\ \text{Angles Formulas}\end{array}$$

$$\sin\theta \cdot 0 - \cos\theta \cdot 1 + \cos\theta \cdot (-1) - \sin\theta \cdot 0 \overset{?}{=} -2\cos\theta \qquad \text{Evaluate each expression.}$$

$$-2\cos\theta = -2\cos\theta \qquad \text{Simplify.}$$

Exercises

Verify that each of the following is an identity.

1. $\sin(90° + \theta) = \cos\theta$

2. $\cos(270° + \theta) = \sin\theta$

3. $\sin\left(\dfrac{2\pi}{3} - \theta\right) + \cos\left(\theta - \dfrac{5\pi}{6}\right) = \sin\theta$

4. $\cos\left(\dfrac{3\pi}{4} + \theta\right) - \sin\left(\theta - \dfrac{\pi}{4}\right) = -\sqrt{2}\sin\theta$

14-6 Study Guide and Intervention

Double-Angle and Half-Angle Formulas

Double-Angle Formulas

Double-Angle Formulas	The following identities hold true for all values of θ. $\sin 2\theta = 2 \sin \theta \cdot \cos \theta$ $\cos 2\theta = \cos^2 \theta - \sin^2 \theta$ $\cos 2\theta = 1 - 2 \sin^2 \theta$ $\cos 2\theta = 2 \cos^2 \theta - 1$

Example   Find the exact values of $\sin 2\theta$ and $\cos 2\theta$ if $\sin \theta = -\dfrac{9}{10}$ and $180° < \theta < 270°$.

First, find the value of $\cos \theta$.

$\cos^2 \theta = 1 - \sin^2 \theta$ $\cos^2 \theta + \sin^2 \theta = 1$

$\cos^2 \theta = 1 - \left(-\dfrac{9}{10}\right)^2$ $\sin \theta = -\dfrac{9}{10}$

$\cos^2 \theta = \dfrac{19}{100}$

$\cos \theta = \pm \dfrac{\sqrt{19}}{10}$

Since θ is in the third quadrant, $\cos \theta$ is negative. Thus $\cos \theta = -\dfrac{\sqrt{19}}{10}$.

To find $\sin 2\theta$, use the identity $\sin 2\theta = 2 \sin \theta \cdot \cos \theta$.

$\sin 2\theta = 2 \sin \theta \cdot \cos \theta$

$\qquad = 2\left(-\dfrac{9}{10}\right)\left(-\dfrac{\sqrt{19}}{10}\right)$

$\qquad = \dfrac{9\sqrt{19}}{50}$

The value of $\sin 2\theta$ is $\dfrac{9\sqrt{19}}{50}$.

To find $\cos 2\theta$, use the identity $\cos 2\theta = 1 - 2 \sin^2 \theta$.

$\cos 2\theta = 1 - 2 \sin^2 \theta$

$\qquad = 1 - 2\left(-\dfrac{9}{10}\right)^2$

$\qquad = -\dfrac{31}{50}$.

The value of $\cos 2\theta$ is $-\dfrac{31}{50}$.

Exercises

Find the exact values of $\sin 2\theta$ and $\cos 2\theta$ for each of the following.

1. $\sin \theta = \dfrac{1}{4}$, $0° < \theta < 90°$

2. $\sin \theta = -\dfrac{1}{8}$, $270° < \theta < 360°$

3. $\cos \theta = -\dfrac{3}{5}$, $180° < \theta < 270°$

4. $\cos \theta = -\dfrac{4}{5}$, $90° < \theta < 180°$

5. $\sin \theta = -\dfrac{3}{5}$, $270° < \theta < 360°$

6. $\cos \theta = -\dfrac{2}{3}$, $90° < \theta < 180°$

14-6 Study Guide and Intervention (continued)

Double-Angle and Half-Angle Formulas

Half-Angle Formulas

Half-Angle Formulas	The following identities hold true for all values of α. $\sin \frac{\alpha}{2} = \pm\sqrt{\dfrac{1 - \cos \alpha}{2}}$ $\qquad$ $\cos \frac{\alpha}{2} = \pm\sqrt{\dfrac{1 + \cos \alpha}{2}}$

Example Find the exact value of $\sin \frac{\alpha}{2}$ if $\sin \alpha = \frac{2}{3}$ and $90° < \alpha < 180°$.

First find $\cos \alpha$.

$\cos^2 \alpha = 1 - \sin^2 \alpha \qquad \cos^2 \alpha + \sin^2 \alpha = 1$

$\cos^2 \alpha = 1 - \left(\frac{2}{3}\right)^2 \qquad \sin \alpha = \frac{2}{3}$

$\cos^2 \alpha = \frac{5}{9} \qquad$ Simplify.

$\cos \alpha = \pm\frac{\sqrt{5}}{3} \qquad$ Take the square root of each side.

Since α is in the second quadrant, $\cos \alpha = -\frac{\sqrt{5}}{3}$.

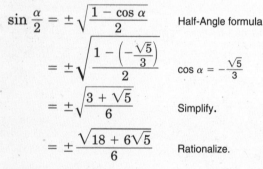

$\sin \frac{\alpha}{2} = \pm\sqrt{\dfrac{1 - \cos \alpha}{2}} \qquad$ Half-Angle formula

$= \pm\sqrt{\dfrac{1 - \left(-\frac{\sqrt{5}}{3}\right)}{2}} \qquad \cos \alpha = -\frac{\sqrt{5}}{3}$

$= \pm\sqrt{\dfrac{3 + \sqrt{5}}{6}} \qquad$ Simplify.

$= \pm\dfrac{\sqrt{18 + 6\sqrt{5}}}{6} \qquad$ Rationalize.

Since α is between $90°$ and $180°$, $\frac{\alpha}{2}$ is between $45°$ and $90°$. Thus $\sin \frac{\alpha}{2}$ is positive and equals $\dfrac{\sqrt{18 + 6\sqrt{5}}}{6}$.

Exercises

Find the exact value of $\sin \frac{\alpha}{2}$ and $\cos \frac{\alpha}{2}$ for each of the following.

1. $\cos \alpha = -\frac{3}{5}$, $180° < \alpha < 270°$
$\qquad\qquad$ **2.** $\cos \alpha = -\frac{4}{5}$, $90° < \alpha < 180°$

3. $\sin \alpha = -\frac{3}{5}$, $270° < \alpha < 360°$
$\qquad\qquad$ **4.** $\cos \alpha = -\frac{2}{3}$, $90° < \alpha < 180°$

Find the exact value of each expression by using the half-angle formulas.

5. $\cos 22\frac{1}{2}°$ $\qquad\qquad$ **6.** $\sin 67.5°$ $\qquad\qquad$ **7.** $\cos \dfrac{7\pi}{8}$

14-7 Study Guide and Intervention

Solving Trigonometric Equations

Solve Trigonometric Equations You can use trigonometric identities to solve trigonometric equations, which are true for only certain values of the variable.

Example 1 Find all solutions of $4 \sin^2 \theta - 1 = 0$ for the interval $0° < \theta < 360°$.

$$4 \sin^2 \theta - 1 = 0$$
$$4 \sin^2 \theta = 1$$
$$\sin^2 \theta = \frac{1}{4}$$
$$\sin \theta = \pm\frac{1}{2}$$
$$\theta = 30°, 150°, 210°, 330°$$

Example 2 Solve $\sin 2\theta + \cos \theta = 0$ for all values of θ. Give your answer in both radians and degrees.

$$\sin 2\theta + \cos \theta = 0$$
$$2 \sin \theta \cos \theta + \cos \theta = 0$$
$$\cos \theta (2 \sin \theta + 1) = 0$$

$\cos \theta = 0$ or $2 \sin \theta + 1 = 0$

$$\sin \theta = -\frac{1}{2}$$

$\theta = 90° + k \cdot 180°;$ $\theta = 210° + k \cdot 360°,$
$\theta = \frac{\pi}{2} + k \cdot \pi$ $330° + k \cdot 360°;$
$$\theta = \frac{7\pi}{6} + k \cdot 2\pi,$$
$$\frac{11\pi}{6} + k \cdot 2\pi$$

Exercises

Find all solutions of each equation for the given interval.

1. $2 \cos^2 \theta + \cos \theta = 1, 0 \le \theta < 2\pi$

2. $\sin^2 \theta \cos^2 \theta = 0, 0 \le \theta < 2\pi$

3. $\cos 2\theta = \frac{\sqrt{3}}{2}, 0° \le \theta < 360°$

4. $2 \sin \theta - \sqrt{3} = 0, 0 \le \theta < 2\pi$

Solve each equation for all values of θ if θ is measured in radians.

5. $4 \sin^2 \theta - 3 = 0$

6. $2 \cos \theta \sin \theta + \cos \theta = 0$

Solve each equation for all values of θ if θ is measured in degrees.

7. $\cos 2\theta + \sin^2 \theta = \frac{1}{2}$

8. $\tan 2\theta = -1$

14-7 Study Guide and Intervention (continued)
Solving Trigonometric Equations

Use Trigonometric Equations

Example LIGHT Snell's law says that $\sin \alpha = 1.33 \sin \beta$, where α is the angle at which a beam of light enters water and β is the angle at which the beam travels through the water. If a beam of light enters water at 42°, at what angle does the light travel through the water?

$\sin \alpha = 1.33 \sin \beta$	Original equation
$\sin 42° = 1.33 \sin \beta$	$\alpha = 42°$
$\sin \beta = \dfrac{\sin 42°}{1.33}$	Divide each side by 1.33.
$\sin \beta \approx 0.5031$	Use a calculator.
$\beta \approx 30.2°$	Take the arcsin of each side.

The light travels through the water at an angle of approximately 30.2°.

Exercises

1. A 6-foot pipe is propped on a 3-foot tall packing crate that sits on level ground. One foot of the pipe extends above the top of the crate and the other end rests on the ground. What angle does the pipe form with the ground?

2. At 1:00 P.M. one afternoon a 180-foot statue casts a shadow that is 85 feet long. Write an equation to find the angle of elevation of the Sun at that time. Find the angle of elevation.

3. A conveyor belt is set up to carry packages from the ground into a window 28 feet above the ground. The angle that the conveyor belt forms with the ground is 35°. How long is the conveyor belt from the ground to the window sill?

SPORTS The distance a golf ball travels can be found using the formula $d = \dfrac{v_0{}^2}{g}\sin 2\theta$, where v_0 is the initial velocity of the ball, g is the acceleration due to gravity (which is 32 feet per second squared), and θ is the angle that the path of the ball makes with the ground.

4. How far will a ball travel hit 90 feet per second at an angle of 55°?

5. If a ball that traveled 300 feet had an initial velocity of 110 feet per second, what angle did the path of the ball make with the ground?

6. Some children set up a teepee in the woods. The poles are 7 feet long from their intersection to their bases, and the children want the distance between the poles to be 4 feet at the base. How wide must the angle be between the poles?